Advances in Transport Network Technologies

Photonic Networks, ATM, and SDH

For a complete listing of the *Artech House Telecommunications Library*, turn to the back of this book.

Advances in Transport Network Technologies

Photonic Networks, ATM, and SDH

Ken-ichi Sato

Artech House
Boston • London

Library of Congress Cataloging-in-Publication Data
Sato, Ken-ichi.
  Advances in transport network technologies: photonic networks, ATM, and SDH/ Ken-ichi Sato.
     p.   cm.
  Includes bibliographical references and index.
  ISBN 0-89006-851-8 (alk. paper)
   1. Data transmission systems. 2. Optical communications. 3. Asynchronous transfer mode. 4. Synchronous digital hierarchy (Data transmission) I. Title.
TK5105.S29   1996
004.6'6–dc20   96-26604
                                                                              CIP

British Library Cataloguing in Publication Data
Sato, Ken-ichi
  Advances in transport network technologies: photonic networks, ATM, and SDH
  1. Telecommunication systems 2. Digital communications
  3. Broadband communication systems
  I. Title
  621.3'82

ISBN 0-89006-851-8

© 1996 ARTECH HOUSE, INC.
685 Canton Street
Norwood, MA 02062

All rights reserved. Printed and bound in the United States of America. No part of this book may be reproduced or utilized in any form or by any means, electronic or mechanical, including photocopying, recording, or by any information storage and retrieval system, without permission in writing from the publisher.
All terms mentioned in this book that are known to be trademarks or service marks have been appropriately capitalized. Artech House cannot attest to the accuracy of this information. Use of a term in this book should not be regarded as affecting the validity of any trademark or service mark.

International Standard Book Number: 0-89006-851-8
Library of Congress Catalog Card Number: 96-26604

10 9 8 7 6 5 4 3 2 1

To
my wife, Fumiko,
and my two children,
Yukiko and Eri

Contents

Preface ... xi

Acknowledgments ... xiii

Chapter 1   Fundamentals of the Transport Network ... 1
     1.1   Transport Network Basics ... 1
         1.1.1   Switched Communications Networks ... 1
         1.1.2   Cross-Connect Functions ... 5
     1.2   Transport Network Hierarchical Structure ... 7
         1.2.1   Transport Network Layered Architecture ... 7
         1.2.2   Path Accommodation Within Transmission Media Network ... 10
         1.2.3   Adaptive Transport Network Control ... 11
     1.3   Transport Network Cost Trend ... 13
     References ... 16

Chapter 2   Synchronous Digital Hierarchy Network ... 17
     2.1   Digital Transport Technology Evolution ... 17
         2.1.1   Plesiochronous Digital Hierarchy (PDH) ... 17
         2.1.2   Synchronous Digital Hierarchy (SDH) ... 18
     2.2   SDH Technologies ... 20
         2.2.1   SDH-Based Transport Network Layers ... 20
         2.2.2   SDH NNI ... 22
         2.2.3   Multiplexing Principles ... 26
     References ... 34

Chapter 3   Asynchronous Transfer Mode Network ... 37
     3.1   Principles of ATM ... 37
         3.1.1   Introduction ... 37
         3.1.2   Current STM-Based Transport Techniques ... 38

		3.1.3	Brief History of ATM	40
		3.1.4	ATM Features	44
	3.2	ATM Transport Technologies		45
		3.2.1	ATM Network Architectural Aspects	45
		3.2.2	ATM Layer Specification	47
		3.2.3	ATM Network Aspects	48
		3.2.4	Virtual Path	51
	3.3	Comparison of ATM and STM		53
		3.3.1	Path Networks and Transport Node Architecture	53
		3.3.2	Path Accommodation	55
		3.3.3	Path Bandwidth Control	59
		3.3.4	Subscriber Networking	67
	3.4	Network Resource Management		68
		3.4.1	Network Resource Management Principle	70
		3.4.2	Network Resource Sharing Principle	82
		3.4.3	CBR Path Accommodation Design	84
		3.4.4	VBR Path Accommodation Design	98
		3.4.5	Evaluations of Network Utilization Enhancement	101
	3.5	Network Element		103
		3.5.1	Cross-Connect System	105
		3.5.2	Other Network Elements	108
	3.6	Conclusion		109
	References			110
Chapter 4		Photonic Transport Network		117
	4.1	Penetration of Photonic Transport Technologies Into Networks		117
		4.1.1	Brief History of Optical Transmission Technology	117
		4.1.2	Penetration of Optical Technologies Into Networks	118
	4.2	B-ISDN Services and Transport Network Requirements		120
		4.2.1	Recent Advances in Transport Technology	120
		4.2.2	Transmission Cost Requirements	121
	4.3	Transport Technology Evolution		123
		4.3.1	Optical Transmission System Technology Advancement	123
		4.3.2	Transport Node Technology Advancement	125
		4.3.3	Transport Network Restoration Technology Advancement	127
		4.3.4	Next Steps to Realize Ubiquitous B-ISDN	127

4.4	Optical Path Technology		129
	4.4.1	Transport Network Photonization	129
	4.4.2	Benefits of WDM	129
	4.4.3	Benefits of Optical Path Technologies	130
	4.4.4	Optical Path Realization Technologies	138
4.5	WP and VWP Accommodation Design		155
	4.5.1	Path Accommodation Design Without Considering Network Restoration	155
	4.5.2	Path Accommodation Design Considering Network Restoration	160
4.6	Optical Path Cross-Connect		175
	4.6.1	Generic Optical Path Cross-Connect System Architecture	175
	4.6.2	Optical Path Cross-Connect Switch Examples	177
	4.6.3	Optical Path Cross-Connect Requirements	185
4.7	Optical Path Economics		194
	4.7.1	Existing Transport Network Cost	194
	4.7.2	Transport Network Cost Reduction With Optical Paths	196
4.8	Conclusion		198
	References		199

List of Acronyms	207
About the Author	213
Index	215

Preface

This book covers recent advances in transport technologies including synchronous digital hierarchy (SDH), asynchronous transfer mode (ATM), and optical network technologies. It provides a picture of recent and future transport network technology trends. This book not only examines the technological details required for network systems development, but also emphasizes the basic and fundamental ideas behind these technologies. This is one of the particular points of this book and is made possible by the experience of the author and his colleagues who have been in the forefront of creating many ATM and photonic network technologies, introducing various new concepts, and developing various related technologies.

This book is structured as follows. First, transport network technologies are explained with an emphasis on the transport network layered architecture. Recently developed transport technologies, including SDH and ATM, are then elaborated. In particular, in a comparison to SDH technologies, the inherent advantages of ATM, and the key technologies needed to fully utilize these advantages, are explored. An emphasis is placed on network aspects, and the key points in creating large scale networks are described. The book then focuses on emerging optical network technologies, which, to date, have been discussed mainly in terms of computer communications rather than from the transport network viewpoint. These technologies are essential to developing the future bandwidth-abundant and ubiquitous B-ISDN.

The author has aimed at presenting the materials so that readers will more easily understand recently developed and cutting-edge transport network technologies. The author has also endeavored to show the motive forces of the new technology developments and what lies behind each technical development. *Advances in Transport Network Technologies: Photonic Networks, ATM, and SDH* will not only serve as an introduction to recently developed technologies but will also provide the readers with further research points for the full development of the technologies. The necessary references are provided. This book is expected to elucidate the technology trends in telecommunications transport networks.

This book is intended for university students in electronics/communications engineering, engineers in the telecommunications industry, network providers and system vendors, and, in particular, researchers in the telecommunications field and communications managers overseeing, designing, or operating digital networks.

Acknowledgments

I prepared this book on weekends and holidays to prevent conflicts with my company work. I would like to offer many thanks to my wife Fumiko and my children Yukiko and Eri for the patience and understanding they extended to me during the many hours I spent preparing this book. Further, I would like to thank many of my colleagues who have provided me with so many stimulating suggestions throughout my career. In particular, I thank Mr. Youichi Sato, Dr. Naoaki Yamanaka, Dr. Hisaya Hadama, Dr. Ryutaro Kawamura, Dr. Satoru Ohta, Dr. Kenji Nakagawa, and Dr. Hitoshi Uematsu for their parts in ATM technologies (Chapter 3) and Dr. Satoru Okamoto, Mr. Yoshiyuki Hamazumi, Mr. Atsushi Watanabe, and Mr. Naohide Nagatsu for their parts in optical network technologies (Chapter 4).

I would also like to express my sincere gratitude to Dr. Tomonori Aoyama, executive manager of NTT Optical Network Systems Labs, for his support and Dr. Sadayasu Ono who encouraged me during the preparation of this book. Finally, I would like to thank Dr. Julie Lancashire for offering me the opportunity to publish this book.

Fundamentals of the Transport Network

This chapter provides the basics necessary for the development of the transport network. First, switching and cross-connect functions are discussed. The transport network layered architecture concept is then introduced, emphasizing the path layer functions that will play an important role in developing a large-scale and flexible transport network. Subsequently, adaptive transport network control techniques that will enhance network flexibility and reliability are explained. Finally, the trend in transport network cost is briefly reviewed.

1.1 TRANSPORT NETWORK BASICS

1.1.1 Switched Communications Networks

Communication across a telecommunications network requires the temporary allocation of network resources. For this purpose, switching systems play a key role in public telecommunications networks. The fundamental switching functions are

- Reducing the number of lines needed to connect all the users accommodated in the network;
- Establishing different connections to allow different combinations of users at different times, thus enabling the network resources to be shared by many users.

Consider N users who want to communicate with each other through a telecommunications network. In the simplest scheme, one connection carrying traffic in both directions is necessary for each user pair. The number of connections required for an N user network is thus $_NC_2 = N(N-1)/2$ as shown in Figure 1.1. The number increases as N increases in a geometric progression. However, if a switch is introduced as shown in Figure 1.2, all users can

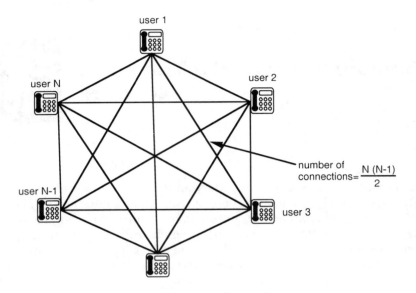

Figure 1.1 Number of network connections among *N* users.

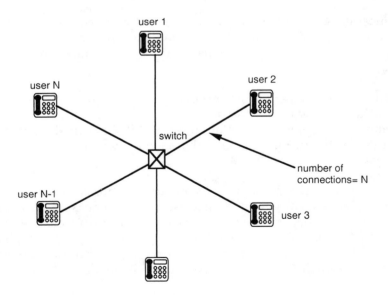

Figure 1.2 Number of network connections among *N* users using a switch.

communicate with each other through a total of N lines. The introduction of a switch enables large savings in the number of connections needed in the network. When N is 1,000, the savings is 99.8%. A network that accommodates a large number of users must, therefore, use switches since the total network cost can be reduced even if switch cost is considered.

The cost of a telecommunications network depends not only on the number of connections in the network, but also on the length of those connections. Figure 1.3 depicts a simple connection network model for two geographically separated user groups. Two different kinds of connections exist: access lines between the users and switches and trunks between the network switches. Usually trunks are longer than access lines. If there are N connections between switches, N users of local network A can always communicate to different N users in local network B at the same time (no blocking). However, to reduce trunk cost, connections between switches are shared by many users, thus minimizing the number of trunk connections. The number of trunk connections required can be determined so that the specific call-blocking probability assigned to the trunk connections is satisfied.

The number of trunk connections can basically be determined from the erlang lost-call formula [1–3] as

$$B(N, A) = \frac{A^N/N!}{\sum_{i=0}^{N} \frac{A^i}{i!}} \qquad (1.1)$$

where $B(N, A)$ is the call-blocking probability, A is the offered traffic intensity (traffic offered in erlangs), N is the available number of circuits, and $N!$ is the factorial N. The numerical calculations are shown in Figure 1.4.

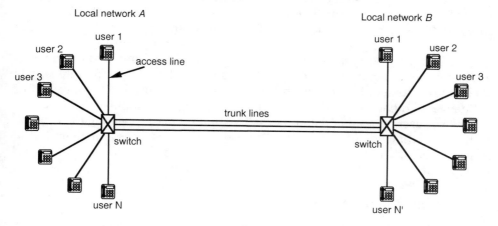

Figure 1.3 Connection network model for two geographically separated user groups.

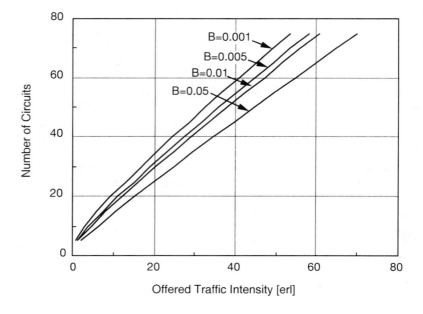

Figure 1.4 Graphical representation of the erlang formula.

It will easily be understood that a single large group of trunk connections has a lower blocking probability than several small ones when the total number of connections is the same for a given traffic flow. In other words, the large group can carry more traffic at a given blocking probability than the total traffic carried by the small groups. For example, when the blocking probability is set to 0.001 (0.1%) and the traffic intensity given to one group is 5 erlangs, the required number of output trunk connections is 14 (see Figure 1.4). Ten such groups can carry 50 erlangs and the total output trunk connection number is 140. However, when the ten groups are combined (the total traffic intensity is 50 erlangs), the required output trunk connection number is 71—only half of the above. In other words, the trunk utilization is almost doubled. Thus, as the number of users accommodated by each switch increases (see Figure 1.3), the utilization efficiency of trunk connections between switches at a given traffic flow increases. This can reduce trunk transmission cost, and the reduction can be significant when the transmission distance is large.

1.1.2 Cross-Connect Functions

As described in the previous section, switches are introduced to develop a large-scale communications network economically, and each switch is connected with trunk connections (bundle of circuits). The trunk connection network is a logical one, and the bandwidth of each trunk is determined according to the traffic demand between switches. Here, the unit of trunk connections is called a path. (It is more widely defined as a grouped circuits serving as a unit of network operation, design, and provisioning.) In the analog networks utilizing frequency division multiplexing techniques, and in the plesiochronous digital hierarchy (PDH) networks, the unit is six telephone channels. To convey paths between switches, physical transmission systems utilizing optical fibers and radio waves are now widely utilized. The path network is fundamentally a logical one and the topology is usually different from that of the physical transmission network, since transmission network topology is determined not only by the traffic demands between switches, but also by considering geographical conditions. Figure 1.5 depicts a traffic network and a transmission facilities network and how they can be realized.

Recent advances in transmission technology can realize very high capacity optical-fiber transmission (i.e., 2.5 Gbps and 10 Gbps). For example, a 2.5-Gbps transmission system can accommodate about 32,000 64-Kbps telephone channels. The traffic between switches (switching units) is not so large,

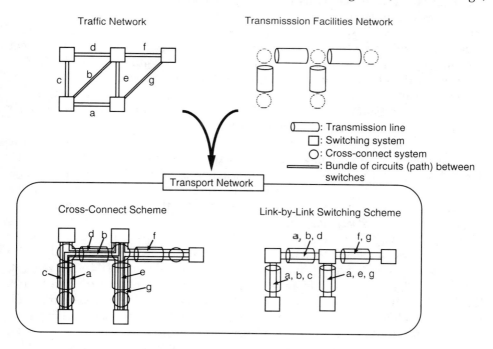

Figure 1.5 Link-by-link switching scheme and cross-connect scheme.

so a large number of paths that have different end points are multiplexed within a transmission line. To accommodate a traffic network (a path network) in a large-capacity transmission facilities network, the cross-connection of the paths [4,5] plays an important role in constructing economical transport networks as explained below.

To accommodate a traffic network, two different schemes can be identified. One is to apply link-by-link switching, and the other is to utilize path cross-connection, as schematically illustrated in Figure 1.5. When link-by-link switching is utilized, link utilization is maximized since total link capacity between switches is shared among many calls that can be accommodated within the link. The path scheme, on the other hand, permits bandwidth sharing among calls only within each path bandwidth. Therefore, as explained in Section 1.1.1, the link utilization becomes higher with link-by-link switching. The increased link utilization efficiency offered by link-by-link switching is offset by an increase in the amount of processing needed. With link-by-link switching, each transit switch has to perform call-by-call processing regarding call admission control that includes call routing and bandwidth allocation. However, with path cross-connection, direct paths are established between two switches, which eliminates call-by-call processing at intermediate cross-connect nodes along paths. Obviously, these two approaches can be combined, with multiple stages of hierarchical switching systems and (multiple stages of) path cross-connect systems. The dramatic cost reductions achieved in transmission cost in the last 15 years with the introduction of optical-fiber transmission systems has not been matched by reductions in processing cost, as demonstrated in Section 1.3. Consequently, recent networks restrict the number of switches used to establish a connection, and switching hierarchies have recently started to flatten [6].

Connection admission control is done on the basis of the usage of paths that have been pre-established between switches considering traffic demands. The role of the path is called service-access. As mentioned before, in the telephone networks using analog and PDH technologies, the path bandwidth for the service-access is six telephone channels. In recently developed synchronous digital hierarchy (SDH) networks, the path bandwidth for service-access is 1.5 Mbps (twenty-four 64-Kbps channels) or 2.0 Mbps (thirty-two 64-Kbps channels). Transmission system bit rates, however, are much larger than the service-access path bandwidths and are now on the order of gigabits per second. Therefore, higher level paths are now utilized in SDH networks. The higher level paths (49 or 150 Mbps) accommodate several service-access paths. The role of the higher level paths, which are accommodated within a higher capacity transmission line, is designated as transaccess within this text. The paths are also utilized as a unit of network restoration, as will be addressed later. The hierarchical path structure is depicted schematically in Figure 1.6.

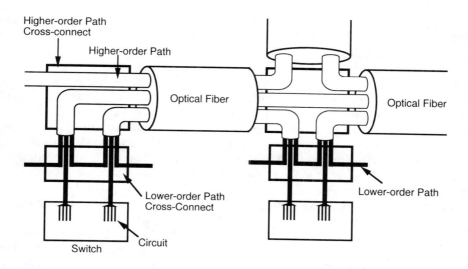

Figure 1.6 Schematic illustration of hierarchical path structure and cross-connect function.

1.2 TRANSPORT NETWORK HIERARCHICAL STRUCTURE

1.2.1 Transport Network Layered Architecture

For the target infrastructure of the future network, reduced network operation cost is an important feature, as is high network reliability. As mentioned above, the capability of recent transmission systems has been increasing with the advancement of optical technology, resulting in increased damage upon single transmission link/system failure. This trend will continue in the future, and hence the development of effective network restoration techniques against such network failures is becoming more and more important. Moreover, the increasing volume of data transmission between computer systems and the increasing value of the data itself urge rapid network restoration to minimize economic loss. Flexibility of the network is another key issue, since it permits a variety of cost-effective end-customer controls and provides the network with adaptability in the face of unknown service requirements.

In this context, a transport network architecture satisfying the above requirements and capable of supporting rather diverse connection demands has to be developed on the basis of network functionality [7], not merely to offer one service. The transport network should be layered according to function. This simplifies the design, development, and operation of the network and allows smooth network evolution. The introduction of the layered concept also makes it easy for each network layer to evolve independently of the other layers by capitalizing on the introduction of new technology specific to each layer. The layering concept has been extensively discussed within ITU-T for the SDH transport network [8].

Figure 1.7 explains a layered network structure for a public telecommunications network. It comprises a service network layer and a transport network layer. The service network layer consists of different service networks, each of which is dedicated to a specific service. This layer provides circuits. A transport network, which is realized with paths and physical transmission media, is less service-dependent. Thus, the telecommunications network can be divided into three layers from the viewpoint of functions: a physical media layer, a transmission path layer, and a communication circuit layer, as shown in Figure 1.7. Operation, administration, and maintenance (OA&M) of these layers is performed by dedicated OA&M systems, and these systems work cooperatively as required (see Figure 1.8).

The circuit is an end-to-end connection established/released dynamically or on the basis of short-term provisioning. The circuit layers are dedicated to specific services such as the public switched telephone service, packetized data communication service, and frame relay service. The transmission media, which interconnects nodes and/or subscribers, is constructed based on long-term provisioning; geographical conditions are taken into consideration. The path layer bridges these two layers and plays an important role in constructing reliable and flexible networks. The path layer provides a common signal trans-

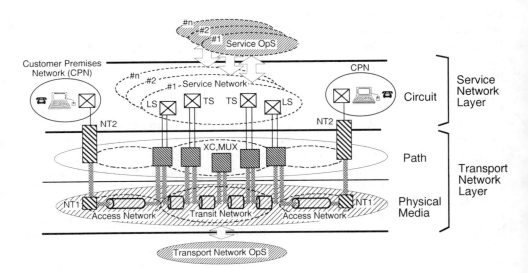

Figure 1.7 Layered architecture of public telecommunications network (*After:* [9]).

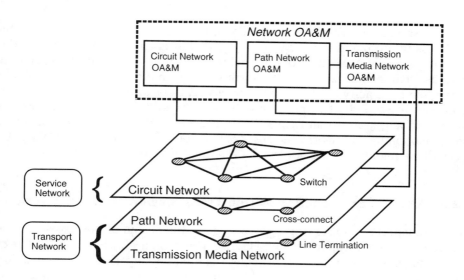

Figure 1.8 Transport network layered architecture and OA&M.

port platform, supporting different service-dependent circuit networks. Functional attributes of a path can be summarized as follows.

- A grouped circuit serving as the unit of network operation, design, and provisioning;
- An object to be manipulated for restoring node and transmission line/system failures.

Effective network restoration (service protection) can be realized in the path layer with the implementation of rerouting or self-healing techniques. Network flexibility can also be enhanced with path layer control. Path layer functions can be performed with intelligent add/drop multiplexer (ADM) systems and/or digital cross-connect systems and the control systems.

Hierarchical Path Network

Digital transmission systems are now being widely introduced into telecommunications networks. The systems have a hierarchical structure in terms of transmission capacities. When a large-capacity digital transmission system accommodates digital paths, the digital paths are required to be cross-connected to construct an economical transport network [10,11].

For service access, the appropriate path bandwidth is determined by the service bandwidth and the traffic demands and is usually not very large; in the existing Japanese SDH network for telephony, it is 1.5 Mbps (twenty-four 64-Kbps channels). Due to recent transmission technology advances, available transmission capacity has been increased to 2.5 Gbps or more. However, traffic demands between any two nodes are not always sufficient to fully utilize the bandwidth, so another path stage for transaccess needs to be introduced to create economical networks as described before; for example, Japan's SDH network for telephony uses 49 Mbps (VC-3) paths as detailed in Chapter 2.

1.2.2 Path Accommodation Within Transmission Media Network

Of the transport network design-related issues, the design of the cross-connection schemes is an important one [11,12]. The design aim is to find the best cross-connection scheme that minimizes the total transmission cost (consisting of the transmission line cost and multiplexer cost). In general, cross-connection at a lower digital path stage results in higher multiplexing cost but low transmission cost since path utilization is enhanced by the finer granularity of the path. On the contrary, cross-connection at a higher digital path stage results in lower multiplexing cost and high transmission cost since each path utilization is reduced due to the coarser granularity of the path [11]. Therefore, to accommodate a bundle of circuits (path) by a transmission facilities network, the multi/demultiplexing stages for cross-connecting, at each intermediate node between end nodes, must be optimized in terms of the total network cost. This is explained in Figure 1.9, where only two digital path stages, D_n and D_{n+1}, are considered for simplicity. In Figure 1.9(a), D_{n+1} digital paths are established between nodes terminating D_n digital paths. Thus each D_n is accommodated within a direct D_{n+1}, and no cross-connection at D_n level is required along each D_n path. This scheme minimizes cross-connection cost for D_n; however, the utilization of the direct D_{n+1} is smaller than that of D_{n+1} shown in Figure 1.9(b). In Figure 1.9(b), D_{n+1} digital paths are established link by link, so each D_n must be cross-connected at every node along the path. This consequently maximizes the cross-connection cost for D_n; however, the utilization of D_{n+1} can be maximized by grooming D_n paths at every node. The scheme that minimizes the total transport cost usually lies between these two extreme cases. The optimization complexity in designing digital paths or cross-connect schemes drastically increases as the number of nodes in the network increases [11].

This complexity stems mainly from the hierarchical path structure and the multiple stages of the multi/demultiplexing needed and often results in lower transmission link capacity utilization. For example, even if the multiplexing efficiency of the lower stage paths into the next stage is 0.8 for each

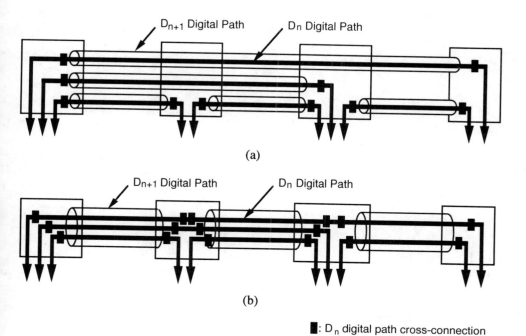

Figure 1.9 Comparison of path establishment schemes with different cross-connects: (a) D_n digital paths accommodated by direct D_{n+1} digital paths and (b) D_n digital paths accommodated by link-by-link D_{n+1} digital paths.

multiplexing stage, three-stage multiplexing results in the final link utilization of only 0.5 (= 0.8^3).

In asynchronous transfer mode (ATM)-based networks, by comparison, virtual path capacity is inherently nonhierarchical and direct multi/demultiplexing of paths into/out of transmission links becomes possible with simple hardware. This greatly simplifies path network architecture and facilitates traffic network accommodation design for multimedia traffic ranging from low-speed data/telephone channels to high-speed data/video channels as depicted in Figure 1.10. The details will be given in Chapter 3.

1.2.3 Adaptive Transport Network Control

The flexibility of a transport network can be enhanced by applying dynamic network controls. Adaptive reconfiguration capability of the network produces many network performance enhancements [5]:

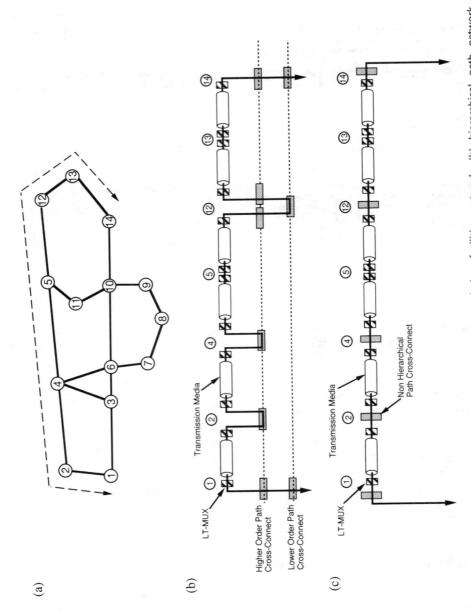

Figure 1.10 Comparison of path accommodation: (a) transmission facilities network, (b) hierarchical path network, and (c) nonhierarchical path network.

- Adaptability to unexpected traffic variations and network failures will be enhanced, thus increasing the reliability of the network.
- The amount of traffic that can be carried in the network will increase.
- The network operating companies can improve the quality of the services they offer—by, for example, offering greater responsiveness to customer demands—and can offer new services on demand at reduced cost.
- Customers will be able to control their own closed networks on a real-time basis—to reconfigure the number of circuits among different locations, for example.

Adaptive control can be applied to both circuit and path layers. For circuit layer control, the means to create adaptable and flexible networks have been studied from different viewpoints. One is the reconfiguration of bundles of circuits employing switches (this method is known as group transit switching [13] or variable communications network [14]). The other is dynamic call routing [15], for which two methods are well known. One predetermines the routing strategy, and changes are made according to the time of the day in order to adapt to the traffic changes caused by the time differences within a country [16]. The other method, the adaptive routing strategy, rapidly follows the dynamic traffic conditions.

As path layer control schemes, dynamic path routing and dynamic path bandwidth allocation are possible through the employment of cross-connect functions [17,18]. A comparison is shown schematically in Figure 1.11(a) of the dynamic connection arrangement of bundles of circuits employing switches, and in Figure 1.11(b) of dynamic path bandwidth allocation employing cross-connect functions. The relation between a traffic network and a facilities network is that depicted in Figure 1.5. Dynamic route arrangement schemes of circuits and paths are not illustrated in Figure 1.11.

In dynamic path layer control, different levels of path management can be identified (Figure 1.12). The processing load required for the control depends on the level as does the utilization of transmission link capacity. This is schematically shown in Figure 1.12. Increasing the link utilization efficiency (decreasing transmission cost) is at the cost of increased processing load (cost). The level of path management to be applied is decided after evaluating the benefits created by the management and determining the ratio of processing and transmission costs. For virtual path bandwidth control in ATM networks, several analytical procedures to determine the processing load and link utilization tradeoffs have been developed [18–20]. They are discussed in Chapter 3.

1.3 TRANSPORT NETWORK COST TREND

The configuration of the telephone network before 1980 was based on relatively low bandwidth transmission systems using copper cables. A large number of

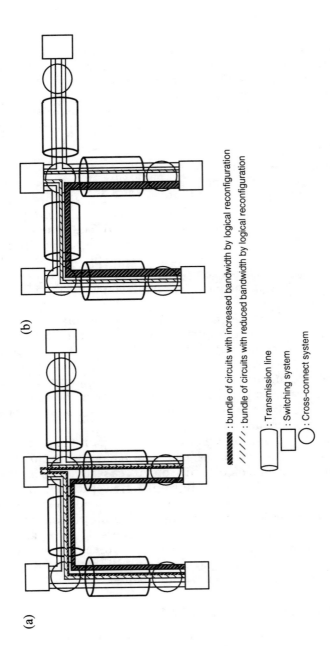

Figure 1.11 Comparison of logical network reconfiguration strategies: (a) dynamic connection arrangement of bundles of circuits employing switches and (b) dynamic path bandwidth allocation employing cross-connect functions.

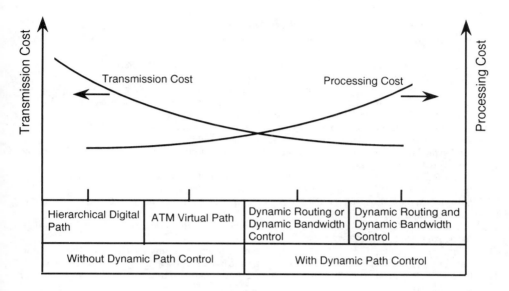

Figure 1.12 Schematic illustration of processing and transmission cost for different levels of path management.

hierarchically organized switching systems were needed to minimize trunk connections or to maximize transmission link utilization in order to reduce the relatively high transmission cost. However, transmission cost has fallen by two to three orders of magnitude over the last 15 years due to the introduction of optical-fiber transmission [21–25]. Meanwhile, cost reduction in switching systems has not been so dramatic [21,22]. This change in the relative costs has led to network designs that require fewer and larger switching systems and fewer switching hierarchies [21–23]. The introduction of optical fibers into subscriber networks enables longer access distances and thus the area covered by a local switching system can be enhanced. In addition, this leads to a reduction in the number of local switching systems and an increase in such systems' capacity.

Future networks will have to carry various nonvoice services. Such services will require much larger bandwidths and have diverse traffic characteristics. To support the services, networks need to be able to transport a large bandwidth at affordable cost and to attain enhanced flexibility. In the following chapters we will examine the recent transport networking technology advances made toward this goal and further discuss future enabling transport networking technologies.

References

[1] Bear, D., *Principles of telecommunication-traffic engineering*, Peter Peregrinus Ltd.: England, 1976.
[2] Ross, M. J., *Fundamentals of digital switching*, New York: Plenum Press, 1983.
[3] Clark, P. M., *Networks and telecommunications*, England: John Wiley & Sons Ltd., 1991.
[4] Power, G., "Seize control with a digital cross-connect," *Telephony*, Aug. 22, 1988, pp. 38–46.
[5] Sato, K., S. Ohta, and I. Tokizawa, "Broadband transport network architecture based on Virtual Paths," *IEEE Trans. on Commun.*, Vol. 38, No. 8, Aug. 1990, pp. 1212–1222.
[6] *IEEE Communications Magazine*, Feature Topic: "The Changing Role of Switching in the Telecommunications Network," Vol. 31, No. 1, Jan. 1993.
[7] Ishikawa, H., "New concept in telecommunications network architecture," *NTT Review*, Vol. 1, No. 1, May 1989, pp. 79–86.
[8] ITU-T Recommendation G.803, Architecture of transport networks based on the synchronous digital hierarchy (SDH), March 1993.
[9] Sato, K., "Transport Network Evolution with Optical Paths," *Proc. ECOC '94*, Sept. 25–29, 1994, Florence, Italy, pp. 919–926.
[10] Stuck, B. W., "The digital cross-connect: Cornerstones of future networks?," *Data Communications*, Aug. 1987, pp. 165–178.
[11] Okano, Y., S. Ohta, and T. Kawata, "Assessment of cross-connect systems in transmission networks," *Proc. GLOBECOM 87*, Tokyo, Japan, Nov. 15–18, 1987.
[12] Okano, Y., T. Kawata, and T. Miki, "Designing digital paths in transmission networks," *Proc. GLOBECOM 86*, Houston, TX, 1986, pp. 25.2.1–25.2.5.
[13] Shimasaki, N., A. Okada, and T. Yamaguchi, "Group transit switching—a new operational approach to be applicable to switched communication network," *Proc. ICC'74*, 11D, June 1974.
[14] Akiyama, M., "Variable communication network design," *Proc. 9th International Teletraffic Congress*, 1979.
[15] Hurley, B. R., C.J.R. Seidl, and W. F. Sewell, "A survey of dynamic routing methods for circuit-switched traffic," *IEEE Communications Magazine*, Vol. 25, No. 9, 1987, pp. 13–21.
[16] Ash, G. R., R. H. Cardwell, and R. P. Murray, "Design and optimization of networks with dynamic routing," *BSTJ*, Vol. 60, No. 10, Oct. 1987, pp. 1787–1820.
[17] Blauer, M., M. Methiwalla, J. Miceli, and J. Yan, "Transport services management," *Telesis*, Vol. 1, 1986, pp. 39–43.
[18] Ohta, S., K. Sato, and I. Tokizawa, "A dynamically controllable ATM transport network based on the Virtual Path concept," *Proc. GLOBECOM '88*, Nov. 28–Dec. 1, Lauderdale, FL, pp. 39.2.1–39.2.5.
[19] Sato, K., and I. Tokizawa, "Flexible asynchronous transfer mode networks utilizing Virtual Paths," Proc. ICC '90, Atlanta, GA, April 16–19, 1990, pp. 318.4.1.–318.4.8.
[20] Hadama, H., K. Sato, and I. Tokizawa, "Dynamic bandwidth control of Virtual Paths in ATM networks," Presented at *3rd IEEE Int. Workshop on Multimedia Commun, MULTIMEDIA '90*, Bordeaux, France, Nov. 14–17, 1990.
[21] White, P. E., "The changing role of switching systems in the telecommunications network," *IEEE Communications Magazine*, Jan. 1993.
[22] White, P. E., "Broadband networking: A new paradigm for communications," Presented at *B-ISDN Networking Workshop at Columbia University*, 1992.
[23] Moondra, S. L., "Impact of emerging switching-transmission cost tradeoffs on future telecommunications network architectures," *IEEE Journal on Selected Areas in Communications*, Vol. 7, No. 8, Oct. 1989, pp. 1207–1218.
[24] Cheung, Nim K., "The infrastructure for gigabit computer networks," *IEEE Communications Magazine*, Vol. 30, No. 4, April 1992, pp. 60–68.
[25] Hagimoto, K., Yukio Kobayashi, and Y. Yamabayashi, "Multi-gigabit-per-second optical transmission systems," *Technical Digest of IOOC '95*, Hong Kong, June 28, 1995, Vol. 2, WC1-1.

Synchronous Digital Hierarchy Network

2

This chapter describes synchronous digital hierarchy (SDH) network technologies. One of the significant transport technologies that appeared in the late 1980s was SDH. The worldwide unique network node interface (NNI) was first standardized in 1988 in CCITT (now ITU-T). SDH systems were deployed extensively in the 1990s in various countries. This yielded the benefits of increased cost-effectiveness and enhanced transmission quality and operation capabilities [1–4]. SDH is thus the foundation on which recent transport networks have been created.

2.1 DIGITAL TRANSPORT TECHNOLOGY EVOLUTION

2.1.1 Plesiochronous Digital Hierarchy (PDH)

Networks installed before 1990 are based on the plesiochronous digital hierarchy (PDH) [5]. In PDH networks, the bit rate of each tributary signal is controlled within a specific limit and is not synchronized with multiplexing equipment. This type of multiplexing is referred to as being *plesiochronous*—that is to say, nearly synchronized. The individual tributaries are synchronized with equipment at each multiplexing step by utilizing positive bit-stuff justification. The asynchronous nature of this multiplexing scheme creates several difficulties. For example, to access a 1.544-Mbps signal for rerouting or testing, it is required that the whole line signal structure be demultiplexed step-by-step down to the 1.544-Mbps level. This entails significant processing overhead and causes inefficiencies. The limitations of PDH are summarized below:

- No direct visibility of tributaries in a multiplexed signal is provided.
- Step-by-step asynchronous multiplexing (positive bit stuffing) and demultiplexing is required.
- Bit multiplexing is employed instead of byte multiplexing.

- Different frame structures are defined at different bit rates.
- Limited network management and maintenance capabilities are possible since relatively few overhead bits are available.
- No worldwide standard exists, so interworking between equipment supplied from different manufacturers cannot be expected.

The digital hierarchies of Europe, North America, and Japan are different, as shown in Figure 2.1.

2.1.2 Synchronous Digital Hierarchy (SDH)

If the network is completely synchronized, direct multiplexing/demultiplexing of transmission signals can be done easily through any network device. Also, any low-speed multiplexing level signals can be accessed so digital cross-connect function and transmission signal monitoring capabilities are more easily implemented. A telecommunications network can therefore be handled as just a single digital circuit pack, and advanced network functions can be realized simply. This background has led CCITT to develop the SDH NNI standards [6–8]. Advantages of SDH are summarized as follows:

- The worldwide standard allows us to create a unified telecommunication network infrastructure, interconnecting network equipment from different vendors. The first level of the synchronous digital hierarchy is 155.520 Mbps and the higher SDH bit rates are defined as integer (N) multiples of the first-level bit rate. At present, the standardized values of N are 1, 4, 16, and 64.
- Direct synchronous multiplexing/demultiplexing is possible; individual tributary signals can be directly multi/demultiplexed into/from higher rate SDH signals, enabling cost-effective and flexible networks.
- Ample overhead bytes (nearly 5% of the total) allow enhanced network OA&M functions.
- SDH can accommodate various existing signal types via virtual containers (VCs). The SDH network can therefore overlay existing networks and flexibly support new service signals.
- The transport network layered concept [9] is effectively realized.

These features are explained in the following sections. Transport flexibility can be enhanced by adopting the ATM techniques, as will be explained in Chapter 3. The ATM techniques (SDH-based ATM) exploit the benefits of SDH; in other words, they retain commonality with SDH (transmission media layer and higher order path layer techniques are common).

Synchronous Digital Hierarchy Network 19

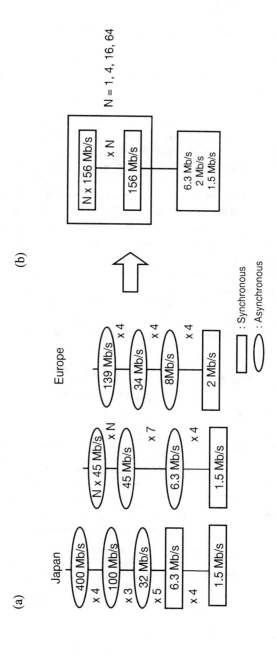

Figure 2.1 (a) PDH and (b) SDH.

2.2 SDH TECHNOLOGIES

2.2.1 SDH-Based Transport Network Layers

Figure 2.2 depicts the layered model of the transport network and an example based on SDH [9]. Different service networks are supported by SDH transport layers. Figure 2.3 shows the layer relationship for the SDH-based transport network [9].

Transmission medium layer networks are dependent on the transmission medium such as optical fiber or radio. Transmission media layer networks are divided into section layer networks and physical media layer networks. Section layer networks are divided into two types: a multiplex section (MS) layer network and a regenerator section (RS) layer network. The multiplex section layer is concerned with the end-to-end transfer of information between locations that route or terminate paths (as shown in Figure 2.4, which shows an SDH network element connection example). The regenerator section layer is concerned with the transfer of information between regenerators and locations that route and terminate paths.

Path layer networks are divided into higher order path layer networks and lower order path networks (Figure 2.3). The higher order paths consist of VC-3 and VC-4 paths, while the lower order paths consist of VC-11, VC-12, VC-2, and VC-3 paths (Figure 2.3). Here, a VC is an information structure that consists of information payload and path overhead (POH). As will be explained

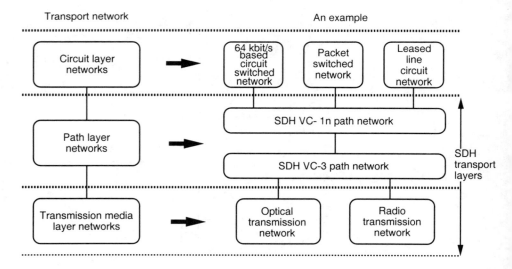

Figure 2.2 Layered model of the transport network.

Synchronous Digital Hierarchy Network   21

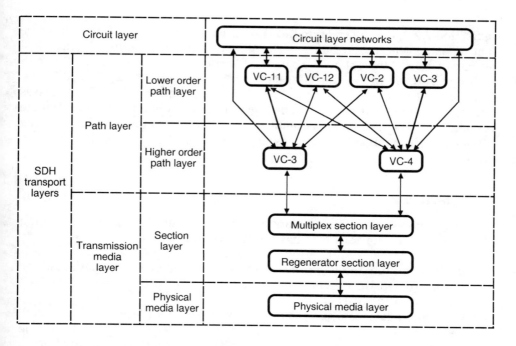

Figure 2.3 SDH-based transport network layered model.

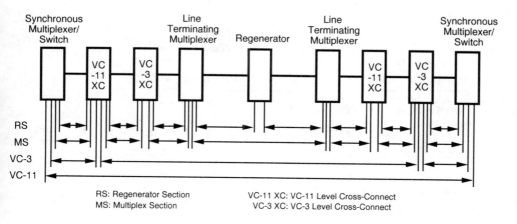

Figure 2.4 SDH network element connection example.

in Section 2.2.3, the information fields are organized in a block frame structure that repeats every 125 or 500 μs. The VC-11 path can contain a hierarchical bit rate of 1.544 Mbps, the VC-12 path a bit rate of 2.048 Mbps, the VC-2 path a bit rate of 6.312 Mbps, the VC-3 path a bit rate of 44.736 and 34.368 Mbps, and the VC-4 path a bit rate of 139.264 Mbps. The lower order VC comprises a single container plus path OH. The higher order VC comprises either a single container-n (n = 3, 4) or an assembly of tributary unit (lower order VC plus tributary unit pointer; [7]) groups together with OH appropriate for the level [10,11].

2.2.2 SDH NNI

Figure 2.5 shows a possible network configuration and illustrates the location of the SDH NNI. The SDH NNI has the following features:

- It is layered into three entities: the section, the higher order path, and the lower order path. Higher and lower order paths are defined as the VCs specified in the SDH NNI.
- It accommodates many types of signals via different VCs (see Figure 2.3).
- Section and path layers have their own overheads for transmitting OA&M information, thus allowing enhanced network OA&M functions.
- It uses a pointer for synchronization. The pointer isolates the VC from the transmission frame or the higher order VC and allows wander and jitter to be accommodated.

In the SDH network, each VC cross-connect function can be easily implemented by utilizing the above features. Moreover, it is possible to independently monitor the performance of section and higher and lower order paths using the overhead bytes, thus enhancing the OA&M capabilities. Since the VCs

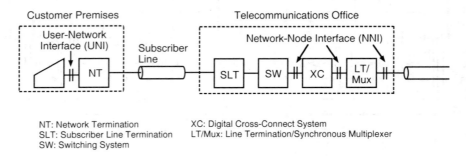

Figure 2.5 Location of NNI.

defined in SDH can carry different bit-rate signals, new services can be flexibly supported. These characteristics are not achieved by PDH networks.

PDH networks have to handle several path stages, such as 1.5 Mbps, 6 Mbps, 32 Mbps, and 100 Mbps. SDH networks have two kinds of paths: higher order and lower order paths. As the path levels that have to be supported in the network decrease, the network becomes more simple, as shown in Figure 2.6. PDH networks need multistage multiplexers (MUXs) to match the hierarchy. In SDH networks, it is necessary to install only lower and higher order digital cross-connect systems to attain high network efficiency. The cross-connect node architecture simplification is shown in Figure 2.7, using Japan's digital hierarchy [12] as an example. This simplification offers increased link utilization (link cost reduction) and transport network node cost reduction. For example, when the average multiplexing efficiency attained by single-stage multiplexing is 0.8, four-stage multiplexing results in the link utilization of 0.4 (= $0.8^{0.4}$).

Table 2.1 summarizes the features and specifications of NTT's synchronous digital transport systems that have been introduced to date [13–16]. Module A is an optical-fiber transmission system based on SDH. It consists of a

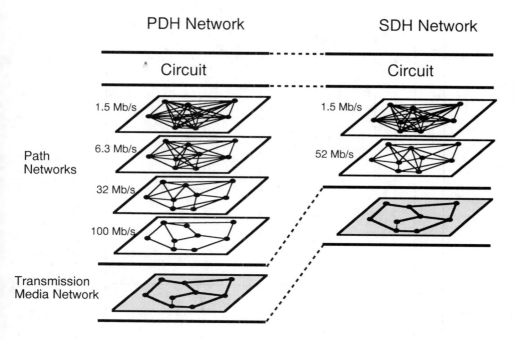

Figure 2.6 Path network simplification.

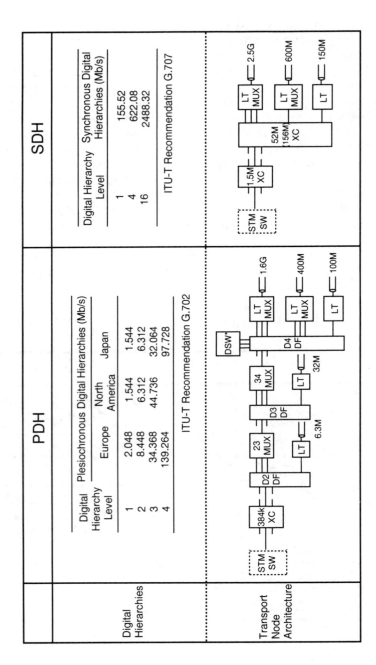

Figure 2.7 Simplification of transport node architecture.

Table 2.1
Synchronous Optical Transport Systems in NTT

Module	Module A				Module B	Module AX	Module C	Module D
Configuration	STM-N → N x 156M OPTICAL LT → MUX → NNI				NNI → (switch)	NNI → (switch)	NNI → MUX → ELI	
Features	- Skip-level multiplexing from intra-office NNI level to STM-N level - Fiber optic line termination - Automatic protection switching - Digital test access - Performance monitoring - Controllable locally or remotely				- VC-11 digital cross-connect - Digital test access - Performance monitoring - Controllable locally or remotely	- VC-3(4) digital cross-connect - Digital test access - Performance monitoring - Controllable locally or remotely	- Skip-level multiplexing from ELI level to intra-office NNI level - Automatic protection switching - Digital test access - Performance monitoring - Controllable locally or remotely	
System	FTM-150M LT MUX	FTM-600M LT MUX	FTM-2.4G LT MUX	FTM-10G LT MUX	XCM-1	XCM-2	TCM-1	TCML-1
Capacity	16,128 CH/Frame (CH: 64 kb/s circuit)	24,192 CH/Frame (CH: 64 kb/s circuit)	32,256 CH/Frame (CH: 64 kb/s circuit)	129,024 CH/F (CH: 64 kb/s circuit)	8,064 x 8,064 P/Switch (P: VC-11 path)	576 x 576 P/Switch (P: VC-3 path)	10,080 CH/Frame (CH:64kb/s circuit)	1,200 CH/Unit (CH:64kb/s circuit)
Interface Rate (Media) — STM-N (inter-office)	155.52 Mb/s (SMF)	622.08 Mb/s (SMF)	2488.32 Mb/s (SMF)	9953.28 Mb/s (SMF)	—	—	—	—
Interface Rate (Media) — Intra-office	155.52 Mb/s 51.84 Mb/s (SMF)	155.52 Mb/s 51.84 Mb/s (SMF)	155.52 Mb/s 51.84 Mb/s (SMF)	51.84, 155.52 622.08, 2,488.32 Mb/s (SMF)	155.52 Mb/s 51.84 Mb/s (SMF)	155.52 Mb/s 51.84 Mb/s (SMF)	155.52 Mb/s 51.84 Mb/s (SMF)	51.84 Mb/s (SMF)
Interface Rate (Media) — ELI (intra-office)	—	—	—	—	—	—	8.192 Mb/s 6.312 Mb/s 2.048 Mb/s 1.544 Mb/s (metallic)	6.312 Mb/s 2.048 Mb/s 1.544 Mb/s (metallic)

Note: SMF: single mode fiber; ELI: existing lower-speed interface.

line-terminating multiplexer and a regenerator. Module B is a VC-11 cross-connect system. It is used for service access and increases the fill factor of higher order paths or transmission line systems. Module AX is a VC-3(4) cross-connect system that is utilized for transmission line access at VC-3(4) level and for path restoration. Module C and D convert the existing lower speed interfaces to the NNI. They have path-terminating functions and perform the signal-mapping functions to and from digital switching systems.

To maximize the utilization of transmission facilities, network engineering, which includes path accommodation design, is essential. In SDH networks, two design steps are needed: one to accommodate lower order paths into higher order paths and another to accommodate higher order paths into a transmission line (section). The rather complicated PDH hierarchical structure of interfaces among component systems and among equipment within a network significantly increases network node cost. Traffic network accommodation design in a transmission facility network, or a transport network design related to cross-connect systems, tends to be very complicated in PDH networks [17,18] because of the many path hierarchical stages. This is apt to deteriorate transmission link utilization efficiency. The larger number of hierarchical stages of paths also makes it difficult to implement dynamic path network reconfiguration capability. In SDH networks, these difficulties have been mitigated.

2.2.3 Multiplexing Principles

Figure 2.8 shows the relationships between various multiplexing elements and possible multiplexing structures. Tributary signals are accommodated within a limited number of standard containers where adaptation functions have been defined for tributaries. A container (C-x) is the information structure that forms the payload for a VC. The following explains the multiplexing elements.

Synchronous Transport Module-N (STM-N)

Synchronous transport module level N (STM-N) is the information structure of level N for SDH and consists of section overhead information fields, administrative unit pointer(s), and information payload, organized in a block frame structure that repeats every 125 μs, as shown in Figure 2.9. The two-dimensional representation for the STM-1 signal frame consists of 9 rows by 270 columns, giving a total signal capacity of 2,430 bytes per 125-μs frame, which corresponds to the bit rate of 155.52 Mbps. There are two reasons for the 9-row structure: A 9-row periodic structure within a 125-μs frame can reduce the memory needed for signal processing in SDH transmission equipment, and the 9-row structure is suitable for accommodating the tributaries of 1.544 Mbps and

Synchronous Digital Hierarchy Network 27

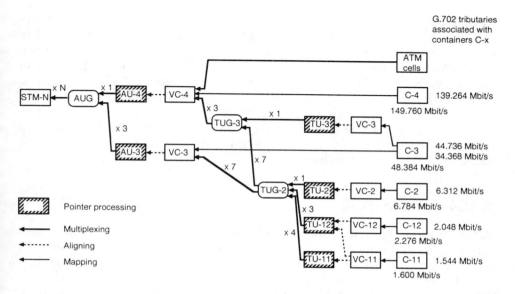

Figure 2.8 Relationship between multiplexing elements and possible multiplexing structures.

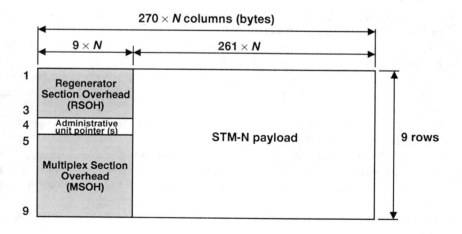

Figure 2.9 STM-*N* frame structure.

2.048 Mbps, since a 1.544-Mbps signal is accommodated within 27 bytes (64 Kbps × 9 rows × 3 columns = 1.728 Mbps) and a 2.048 Mbps signal is accommodated within 36 bytes (64 Kbps × 9 rows × 4 columns = 2.304 Mbps).

STM-1 consists of a payload that contains one VC-4 or three VC-3's, a section overhead (SOH), and administrative unit pointer(s). An STM-N (N = 4, 16, and 64) signal is assembled by byte interleaving N parallel frame-synchronized STM-1 payloads and AU pointers (see below). The first 9 × N columns are occupied by SOH and administrative unit (AU) pointers. The remaining columns are occupied by the N × (VC-4 or three VC-3) signals associated with the N individual STM-1 signals.

Section Overhead (SOH)

SOH contains frame synchronization signals and OA&M-related signals. It pertains only to an individual transport system and is not transferred with the VC between transport systems. SOH is divided into regenerator section overhead (RSOH) (rows 1–3) and multiplex section overhead (MSOH) (rows 5–9), as shown in Figure 2.10 for STM-1. The function of each overhead byte is as follows.

- A1 and A2, for framing purposes. The 16-bit frame alignment word formed by the last A1 byte and the adjacent A2 byte in the transmitted sequence uniquely define the frame reference for each signal rate.
- C1, the STM identifier byte. C1 is used to uniquely identify each interleaved STM in an STM-N signal. It takes a binary value equivalent to the position in the interleave. The C1 in STM #1 takes value 0000 0001 while the C1 in STM #16 takes value 0001 0000.
- B1, for regenerator section error monitoring. The mechanism is bit-interleaved parity 8 (BIP-8).
- E1, for the regenerator section engineering order wire. A channel for voice communication between maintenance personnel at terminal and/or regeneration sites.
- F1, for the user channel. Typically used by operators for remotely recognizing various physical alarms.
- D1, D2, D3, for the data communication channel (DCC). A channel of 192-Kbps capacity used for messaging to and between regenerators. It is typically used for intersystem communications for management and supervision of regenerators.
- B2, for multiplex section error monitoring. BIP 24 × N error-monitoring code.
- K1, K2, for the automatic protection switching channels.

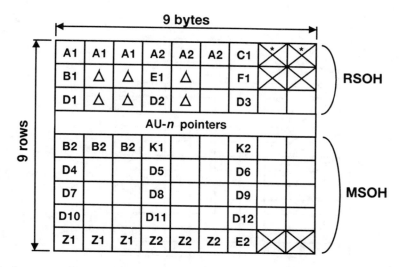

Figure 2.10 STM-1 section overhead.

- D4–D12, for the multiplex section DCC. The 576-Kbps channel for messaging between MS trail terminations in adjacent network nodes.
- Z1. Z2, reserved for future use.
- E2, for the MS engineering order wire. A 64-Kbps channel for voice communication between MS trail terminations in network nodes.

Some overhead bytes of the STM-4 overhead are identical in number to those of STM-1, but others are quadrupled from those of STM-1, as shown in Figure 2.11.

Administrative Unit-n (AU-n)

An administrative unit consists of a higher order VC and an administrative unit pointer, which indicates the offset of the payload frame start from the multiplex section frame start. Two administrative units are defined: AU-4 and AU-3. AU-4(3) consists of a VC-4(3) plus an administrative unit pointer, which indicates the phase alignment of the VC-4(3) with respect to the STM-N frame.

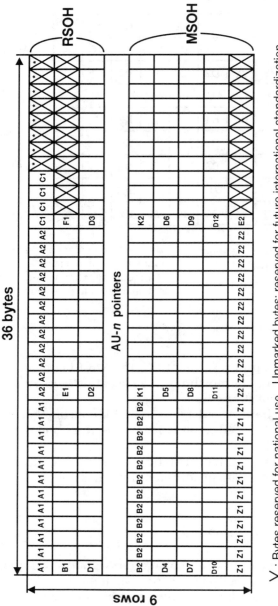

Figure 2.11 STM-4 section overhead.

The VC-n associated with each AU-n does not have a fixed phase with respect to the STM-N frame. The AU-n pointer indicates the location of the first byte of the VC-n and is located in a fixed location (row 4, see Figure 2.9) in the STM-N frame.

AU-n Pointer

The AU-n pointer provides a method that allows flexible and dynamic alignment of the VC-n within the AU-n frame. The VC-n is allowed to "float" within the AU-n frame. This means that VC-n may begin anywhere within the AU-n payload. Usually, the VC-n begins in one AU-n frame and ends in the next. This mechanism minimizes the processing delay needed for multiplexing. Generally, the frame phases transferred to a node from different nodes do not coincide with each other due to differing amounts of transmission delay. In order to align the phase of different signal frames, frame memory is utilized, and, therefore, delays of half the frame time are incurred on average (one frame time maximum). This method demands large memories if the transmission speed increases, which increases the power consumption of the system and hinders system downsizing. The pointer technology resolves these problems. It was first proposed in SONET [19] and then adopted as an ITU-T standard.

The AU-4 pointer is contained in the three bytes of H1, H2, and H3, as shown in Figure 2.12. The pointer contained in H1 and H2 indicates the location of the first byte (J1) of the VC-n. The AU-4 pointer value ranges from 0 to 782, which indicates the offset in 3-byte increments between the pointer and the first byte of the VC-4.

When there is a frequency offset between the frame rate of the AUG (administrative unit group; a homogeneous assembly of AU-3s or an AU-4) and that of the VC-n, the pointer value will be changed as needed, and pulse-stuff justification is employed using positive and negative justification bytes as shown in Figure 2.12.

AU-4 Concatenation

In order to offer greater payload capacity than the 149.76 Mbps provided by VC-4, AU-4s can be concatenated to form an AU-4-Xc. A concatenation indication is contained in the AU-4 pointer and shows that the multi C-4 payload being carried in a single VC-4-Xc should be treated as a single entity through the network for multiplexing and cross-connection. The VC-4-Xc consists of path overhead (the first column) and a single container capable of carrying a tributary signal of $149.76 \times X$ Mbps ($64k \times 9 \times 260 \times X$; 599.040 Mbps for $X = 4$ and 2,396.160 Mbps for $X = 16$), and fixed stuff bytes ($X - 1$ columns)

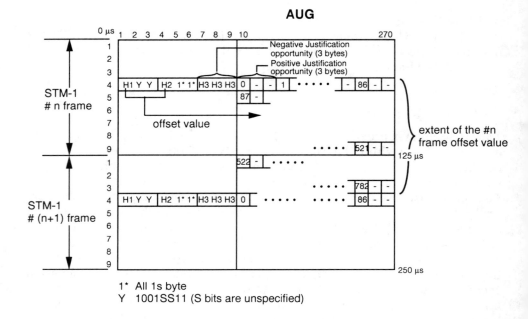

Figure 2.12 AU-4 pointer offset numbering.

as shown in Figure 2.13. The fixed stuff bytes occupy columns 2 to X of the VC-4-Xc.

Mapping of ATM Cells

As the physical layer technology of the ATM transport network, two options have been proposed: cell-based and SDH-based ATM transport networks [20]. Detailed standards have been issued for the SDH-based ATM transport network [8,21]. This is because exploiting already developed SDH technologies realizes the economical introduction of ATM techniques with minimum delay and maximum efficiency through the utilization of SDH network operation capabilities. Thus, ATM networks can be introduced by overlaying SDH networks.

VC-4 and VC-4-Xc are utilized to convey ATM cells (see Chapter 3). The ATM cells (53 bytes) are mapped into a C-4 (C-4-Xc) with its octet boundaries aligned with the C-4 (C-4-Xc) byte boundaries. The C-4 (C-4-Xc) is then mapped into VC-4 (VC-4-Xc) together with the VC-4 OH (VC-4-Xc OH and $(X - 1)$ columns of fixed stuff), as shown in Figures 2.14 and Figure 2.15. The ATM cells may cross a C-4 (C-4-Xc) boundary since no integer multiple of cells matches the C-4 (C-4-Xc) capacity (44.1509......cells/C-4).

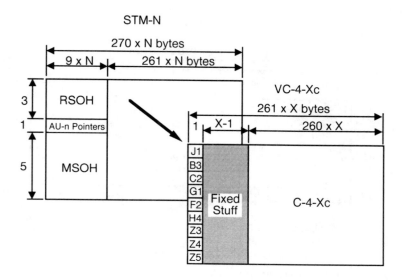

Figure 2.13 VC-4-*X*c structure.

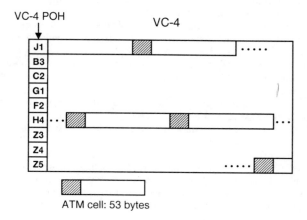

Figure 2.14 Mapping ATM cells in the VC-4.

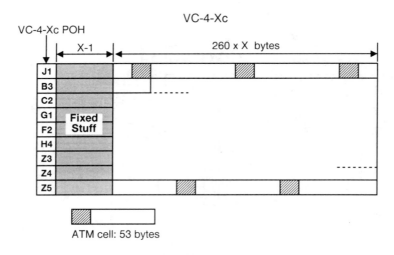

Figure 2.15 Mapping ATM cells into the VC-4-Xc.

References

[1] *IEEE Communications Magazine*, "Global Deployment of SDH-Compliant Networks," Vol. 28, No. 8, Aug. 1990.
[2] *IEEE LTS*, Feature Topic: SONET: No Longer Just a Concept, Vol. 2, No. 4, Nov. 1991.
[3] *IEEE Communications Magazine*, Feature Topic: SDH/SONET, Vol. 31, No. 9, Sept. 1993.
[4] Cannon, H., S. Chum, "International gateway for SDH and SONET interconnection," *GLOBECOM '94*, San Francisco, Nov. 28–Dec. 2, 1994, pp. 72515–18734.
[5] CCITT Blue Book, Volume III—Fascicle III.4, General Aspects of Digital Transmission Systems; Terminal Equipments, Recommendations G.700-G.795, Nov. 1988.
[6] ITU-T Recommendation G.707, Synchronous digital hierarchy bit rates, March 1993.
[7] ITU-T Recommendation G.708, Network node interface for the synchronous digital hierarchy, March 1993.
[8] ITU-T Recommendation G.709, Synchronous multiplexing structure, March 1993.
[9] ITU-T Recommendation G.803, Architectures of transport networks based of the synchronous digital hierarchy (SDH), March 1993.
[10] Sexton, M., and A. Reid, *Transmission Networking: SONET and the Synchronous Digital Hierarchy*, Norwood, MA: Artech House, 1992.
[11] Lee, B. G., M. Kang, and J. Lee, *Broadband Telecommunications Technology*, Norwood, MA: Artech House, 1993.
[12] Ueda, H., H. Tsuji, and T. Tsuboi, "New synchronous digital transmission system with network node interface," *Proc. Globecom '89*, Dallas, Nov. 27–30, 1989, 42.4.1–42.4.5.
[13] Kasai, H., T. Murqase, and H. Ueda, "Synchronous digital transmission systems based CCITT SDH standard," *IEEE Communications Magazine*, Aug. 1990, pp. 50–59.
[14] Miura, H., K. Maki, and K. Nishihata, "SDH network evolution in Japan," *IEEE Communications Magazine*, Feb. 1995, pp. 86–92.
[15] Miura, H., "The fundamentals of link system construction," *NTT Review*, Vol. 6, No. 6, 1994, pp. 16–23.

[16] Nishihata, K., "Trunk link systems—Phase II SDH transmission system," *NTT Review*, Vol. 6, No. 6, 1994, pp. 24–32.
[17] Okano, Y., T. Kawata, and T. Miki, "Designing digital paths in transmission networks," *Proc. GLOBECOM 86*, Houston, TX, 1986, pp. 25.2.1–25.2.5.
[18] Okano, Y., S. Ohta, and T. Kawata, "Assessment of cross-connect systems in transmission networks," *Proc. GLOBECOM 87*, Tokyo, Japan, Nov. 15–18, 1987.
[19] Boehm, R. J., Y. C. Ching, and R. C. Sherman, "SONET (Synchronous Optical Network)," *Proc. GLOBECOM '85*, 1985, 46.8.1.
[20] ITU-T Recommendation I.311, B-ISDN general network aspects, March 1993.
[21] ITU-T Recommendation I.432, B-ISDN user-network interface-physical layer specification, March 1993.

Asynchronous Transfer Mode Network 3

This chapter describes asynchronous transfer mode (ATM) network techniques. They have been extensively developed worldwide since the early 1990s, and their development is still in progress. They are recognized to be very suitable for creating the multimedia communications networks that will be the foundation on which the information society will be based. In this chapter, we first identify current synchronous transfer mode (STM)-based network problems in terms of multimedia transmission network development. The technological points of ATM are then described. The important concept of virtual path (VP), which makes the best use of inherent ATM capabilities and plays a key role in realizing a powerful transport network, is then elaborated. It will be demonstrated how VPs will enhance network performance and capabilities. Network management techniques that are vital to design ATM networks and to reap all the benefits of ATM are also described. ATM network hardware systems are illustrated.

3.1 PRINCIPLES OF ATM

3.1.1 Introduction

In this decade, the need for broadband ISDN (B-ISDN) has been recognized, and great effort is currently being focused on the development of broadband transport technology, including information transport techniques, transport systems architecture, user network interfaces (UNI), network node interfaces (NNI) [1,2], and network introduction strategies. The future telecommunications network should provide various information services in the most efficient and economical way possible [3,4]. The information bit rates transported in the network will range from one thousand to hundreds of millions of bits per second, and the statistical nature of each type of service will differ. This wide range of bit rates and different statistical natures have to be handled efficiently by the network. The network should also be able to evolve to meet unknown future demands for increased flexibility, capacity, and reliability.

Transport techniques based on ATM are being recognized as effective in constructing a network that reflects the features mentioned above. ATM systems have become possible through recent technological innovations such as fiber optics, which enable high-quality (low bit error rate) transmission, large-scale integrated circuits (LSIs), and microprocessors that enable high-speed processing at transport nodes. In ATM, information is organized into short fixed-length information blocks (payload) to which are attached short cell identification labels (header). The labeled blocks (called cells) are transported to their destination according to the routing labels in each cell header (see Figure 3.1). By making the most of the burstiness of information flows and by employing the store-and-forward process for transportation (which introduces variable cell delay into the network), ATM can economically provide integrated transport capability at virtually any bit rate. In an ATM-based network, as will be described later, hierarchical structures of channel/path (bundles of circuits) and the resulting rather complicated TDM frame structure of interfaces can be eliminated. Therefore, the multiplexing and transport aspects of ATM greatly impact network architecture.

Recently, very large capacity optical transmission systems [5–7] (such as 2.5-Gbps and 10-Gbps systems) have been developed and are now in practical use. The latest fiber-optic transmission technologies have significantly reduced the transmission cost portion of the total network cost. The trend toward reduced transmission cost is expected to continue into the future, thus making the node cost relatively high. This leads to the conclusion that total network cost would be reduced most effectively by node cost reduction, which could be achieved by simplifying both node functions and transport network architecture. ATM techniques are discussed against the existing trends in network technology.

3.1.2 Current STM-Based Transport Techniques

Current STM-based network systems are dedicated to specific services, such as plain old telephone service (POTS), circuit and packet-switching services

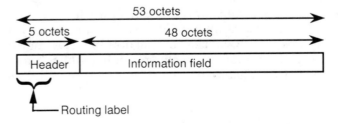

Figure 3.1 ATM cell.

for low-speed data (less than 64 Kbps), narrowband ISDN services, broadband switching services for video communication, and different kinds of leased line services. Because current networks were designed for specific applications, they are difficult to adapt to new services. Current network systems are schematically illustrated in Figure 3.2. The inherent problems of STM-based networks have been recognized [8], some of which are described below from the viewpoint of information transport techniques.

STM-based networks are based on time-division multiplexing (TDM) frame structures. PDH systems were developed with the least overhead to efficiently multiplex transmission channels and paths in a hierarchical manner. The rather complicated and inflexible TDM frame structure of interfaces [9,10] among component systems and among equipment within each network has been significantly improved by the adoption of SDH techniques as discussed

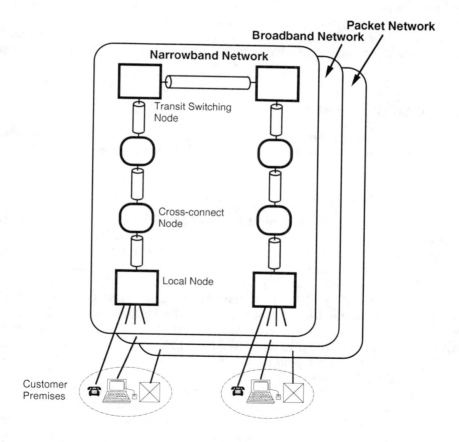

Figure 3.2 Schematic illustration of current transport network.

in Chapter 2. However, inefficiency and inflexibility remain to prevent it from fully supporting a variety of services with different information transfer rates. Offering multimedia services over STM networks significantly increases network node costs. For multiplexing/demultiplexing service channels into/out of the transmission lines, existing STM facilities require multiple multi/demultiplexing processing stages, which deteriorates the utilization of transmission line capacity [11] and necessitates a rather complicated design for accommodating traffic networks within the transmission facilities network [11,12]. For example, even if the utilization efficiency of paths at each stage is 0.8, three-stage multiplexing results in the final link utilization of only 0.5 (= 0.8^3). These problems become more and more serious as demand increases for a variety of services with a wide range of information transfer rates. Furthermore, the multiple multiplexing/demultiplexing systems needed and the hierarchical path structure prevent flexible logical network reconfiguration such as dynamic path routing and dynamic path bandwidth allocation. This also imposes significant restrictions on end-customer control of closed networks. ATM was proposed to overcome the problems of STM and achieve truly effective multimedia communication.

3.1.3 Brief History of ATM

Before going on to the details of ATM, let's briefly look at other technologies from the viewpoint of broadband multimedia information transport capability.

Packet Switching

Packet switching was first proposed in 1964 [13] as a means of voice and data communications for military systems that require high survivability. This is possible because packet switching does not need centralized network components. In packet switching, the bandwidth is dynamically allocated on a link-by-link basis rather than allocating bandwidth over the whole source to destination connection. Data are transferred as a series of packets. In 1967, ARPANET was planned as a pilot project in the United States to link time-shared computer systems. It was the first packet-switching network to serve a large number of distributed users. ARPANET proved that dynamic allocation techniques such as packet switching can be successfully implemented for data communications to share computer resources [14].

Each packet, which has a variable but limited size, includes an information field and a header field, which contains information that determines how the packet will be routed and processed at each intermediate node along the way. Thus, long messages are divided into multiple packets for transmission through the network and reassembled again at the destination terminal. Within the digital pipe (called a *path* hereafter) defined in the network, packets from different users are multiplexed. Packet multiplexing introduces a statistically varying amount of delay for each packet due to packet buffering when the incoming data rate exceeds the available outgoing path/facility data rate.

In contrast, circuit-switching techniques divide the transmission bandwidth into fixed-rate channels. Packet switching eliminates the need to fix channel rates and so is far more flexible in accepting arbitrary data rates, up to the maximum path or transmission system rate. This is possible by choosing the packet size and the frequency with which packets are sent.

The advantages of packet switching are partially offset by the many switches and processors needed in the network and the complex control procedures employed by nodes. Therefore, the available packet transmission speed was limited to less than a few megabits per second. For the user-network interface, the X.25 standard for packet switching was developed in CCITT [15]. The X.25 standard is a user-network interface, so it does not constitute a switch-to-switch protocol. In the network, packets can be transferred on a virtual circuit basis or datagram basis [16].

Fast Packet Switching

Fast packet switching (FPS) [17–20] is based on the basic techniques of packet switching. It adopts simplified protocols and routing strategies. FPS can be connection oriented—connection can be established in the call setup phase. Packets are routed on the route established, and the switches in the network do not correct for packets that are lost or damaged due to transmission error or buffer overflow at multiplexers/switches. This elimination of correction is made possible by the excellent bit error rate performance of recently introduced optical-fiber transmission systems. The packet overhead can be minimized accordingly. The buffering delays can be minimized if the packet length is short and the outgoing path/facility bit rate is high. Thus, FPS minimizes the queuing delay and delay variations. In FPS and fast circuit switching (FCS), delay and/or information loss can occur at network nodes. FPS, however, spreads delay and loss more evenly over all connections, minimizing the impact on any one call. With FCS, the effect tends to be absorbed by one or a few connections, resulting in more severe delay and/or loss [3].

Burst Switching

Burst switching was proposed by the researchers at GTE [21–23] in 1983. A burst is a talk spurt of data messages. Burst switches employ silence detection of speech so that transmission channels are dedicated only while information exists, that is, during talk spurts. A burst begins with a 4-byte header containing the destination address; an identifier of the kind of bursts such as voice, data, or command burst; and a header checksum that prevents delivery of the burst to a wrong destination. The burst ends with an end of message byte.

The links between burst switches are T1 spans (1.544 Mbps), though other rates could be used. Bursts are sent between switches in the TDM channels of the span, with each succeeding character of the burst in succeeding frames of the span. The character rate within the channel for T-carrier is 8 Kbps, the same as the speech codecs' character generation rate. Therefore, no speed buffering is required at link switches for voice bursts. On the other hand, in packet switching, a packet is transmitted between nodes using a high-speed path/link. Thus, a packet's characters will be accumulated at the source rate, and then transmitted at a higher rate. In burst switching, therefore, such packetizing delays do not exist and there is no limitation on voice burst size, resulting in significantly smaller overhead inefficiency.

Channel congestion occurs if there are more bursts on a link's queues than there are idle channels on the links. While a burst awaits assignment to an idle output channel, input continues to the buffer. If 2-ms worth of voice samples accumulate and output has not begun, the accumulated samples are discarded. This is called clipping.

Burst switching is a hybrid of packet and circuit switching. So, limited flexibility is available for a multimedia environment in which voice traffic will not be dominant, although the number of the connections may be large. This technique, therefore, was not adopted as the basis of B-ISDN.

Asynchronous Time Division (ATD)

In the early 1980s, a new technique called asynchronous time divison (ATD) was proposed by researchers at CNET [24–27]. The concept was proposed as a switching and multiplexing technique for a universal and adaptable digital network able to integrate all types of services. The technique can be seen as being intermediate between the packet-switching technique and the circuit-switching technique. Information to be sent is broken into fixed-length blocks (for the first experimental system, called PRELUDE, it was 128 bits comprising a 1-byte label [26]) and multiplexed with blocks from other information sources on the framed transmission carrier as shown in Figure 3.3. In multiplexing, the time slots are not preassigned, and specific labels to identify connections are used. Thus, any bit rate can be transported—each connection uses just the resources needed depending on its real information bit rate. All the blocks in a connection follow the same route as in the case of virtual circuit mode [16] in

Asynchronous Transfer Mode Network 43

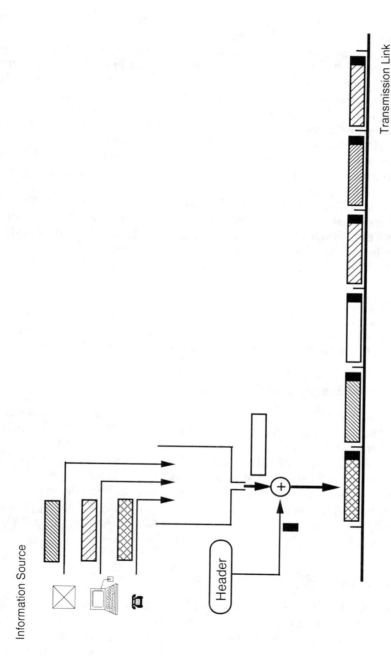

Figure 3.3 Schematic illustration of multiplexing information blocks from different information sources.

packet switching. The switching and multiplexing mechanisms are maximally simplified so that high throughput can be attained.

The transmission resources are used asynchronously, and this is made possible by the use of buffer functions in multiplexers and switches. This incurs switching/multiplexing delay and delay distribution, the degrees of which depend on block size (it is approximately proportional to the block size). These delays increase as the number of switching/multiplexing nodes traversed increase. Therefore, to reduce delay and delay distribution, block size should be as small as possible. This, however, decreases transmission efficiency due to the fixed-length overhead bytes. For supporting circuit-mode connections, the delay distribution has to be smoothed at the reception terminal to provide a consent bit rate.

This ATD technique was first proposed to CCITT (now ITU-T) at the Kyoto Meeting in 1985 as a new transfer mode for B-ISDN [28,29]. International discussion of the B-ISDN transfer mode began, and the subsequent worldwide intensive effort for the development and standardization of ATM [30] was initiated. Thus, the ATD technique provided the basis for ATM.

3.1.4 ATM Features

Before going on to the technical details of ATM, its outstanding features as regards transport network development are briefly summarized below.

Inherent Rate Adaptation and Flexible Interface Structure

ATM provides integrated communications capability by offering constant and variable bit rate channels at virtually any bit rate. This is provided without a complicated interface structure, which carries a multiplicity of fixed channel rates. Customers can dynamically fashion virtual channels (see Section 3.2) to meet their immediate service needs within a simple interface structure. No modifications of the interface structure are required to provide new services. Thus, ATM provides basic communications capabilities in an application-independent fashion and can be used as a foundation that supports a variety of applications.

Nonhierarchical Multiplexing

The information block (cell [31]) is independent of the channel rates and enables direct multiplexing/demultiplexing of service channels/paths that have a variety of bit rates into/out of the transmission lines using simplified hardware and software [32]. Thus, the multiple stages of multiplexers/demultiplexers needed by existing facilities are eliminated. Capitalizing on this nonhierarchical multiplexing capability, logical reconfiguration of the network, such as dynamic bandwidth allocation and routing of paths, can be implemented in a very effective way, as will be discussed in Section 3.3.

3.2 ATM TRANSPORT TECHNOLOGIES

3.2.1 ATM Network Architectural Aspects

Network Layering [33]

The ATM transport network consists of two layers, the ATM layer and the physical layer, as shown in Figure 3.4. The ATM layer is divided into two levels: the virtual channel (VC) level and the virtual path (VP) level. The physical layer is divided into three levels: the transmission path level, the digital section level, and the regenerator section level. Figure 3.5 shows the hierarchical level-to-level relationship in the ATM transport network. Each level relationship uses four architectural components, namely, connection endpoint, connecting point, connection, and link. Here, connection endpoint provides connection termination functions. The connecting point is located inside a connection where two adjacent links come together and provides the connection function. The connection provides information transfer capability between connection endpoints. As shown in Figure 3.5, a connection at a specific level provides services to a link in the next higher level. The link provides the capability for

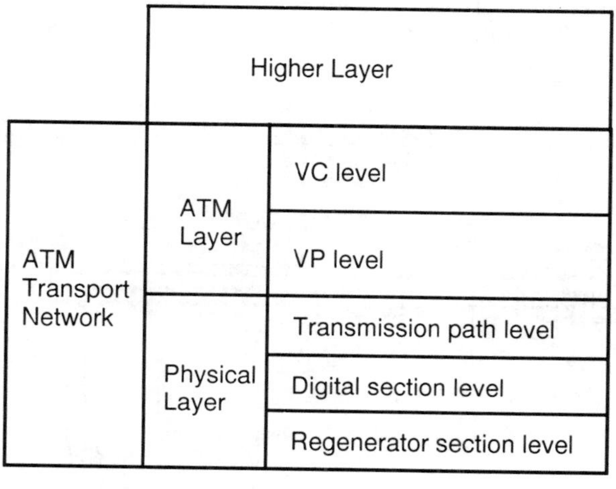

		Higher Layer	
ATM Transport Network	ATM Layer	VC level	
		VP level	
	Physical Layer	Transmission path level	
		Digital section level	
		Regenerator section level	

Figure 3.4 Hierarchy of the ATM transport network.

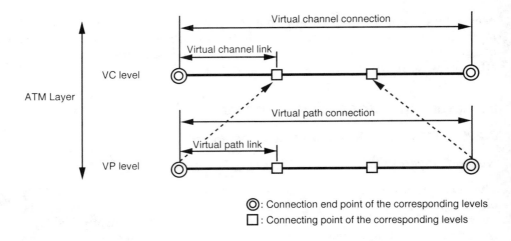

Figure 3.5 Hierarchical level-to-level relationship.

transferring information transparently. Figure 3.6 shows an example of an ATM network structure and corresponding ATM transport network layer functions. The VP and VC play similar roles to the digital path and circuit, respectively, in STM networks.

VC is a generic term used to describe a unidirectional communication capability for the transport of ATM cells. A virtual channel identifier (VCI) included in each cell's header identifies a particular VC link for a given virtual path connection (VPC). Thus, VCI values used in one VP are also used in other VPs. A specific VCI value is assigned each time a VC is switched at a VC

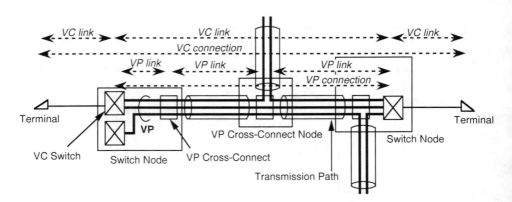

Figure 3.6 An example of ATM network configuration and the architectural components.

connecting point (VC switch) according to the VPI/VCI translation table, which was set up in the connection establishment phase, and the cells are transferred to the next VC connecting point (VC switch). A VC link is a unidirectional capability for the transport of ATM cells between two consecutive ATM entities where the VCI value is translated. A VC link is originated or terminated by the assignment or removal of the VCI value. VC links are concatenated to form a virtual channel connection (VCC). A VCC extends between two VCC endpoints or, in the case of point-to-multipoint arrangements, more than two VCC endpoints. A VCC endpoint is the point at which the cell information field is exchanged between the ATM layer and the user of the ATM layer service. At the VC level, VCCs are provided for information transfer between user and user, user and network, and network and network.

VP is a generic term for a bundle of virtual channel links that have the same endpoints. The virtual path identifier (VPI) included in each cell's header identifies each VP in one transmission line. Thus, VPI values used in one transmission line are also used in other transmission lines. A specific value of VPI is assigned each time a VP is switched or cross-connected in the network. When translating VPI/VCI values as directed by management plane functions and not by control plane functions, the network element is called a VPI/VCI cross-connect. When the network element translates VPI/VCI values under the control of control plane functions, it is called a VPI/VCI switch. A VP link is a unidirectional capability for the transport of ATM cells between two consecutive ATM entities where the VPI value is translated. A VP link is originated or terminated by the assignment or removal of the VPI value. Routing functions for VPs are performed at VP cross-connects (or switches). This routing involves translation of the VPI values of the incoming VP links into the VPI values of the outgoing VP links. VCI values of each VP are not translated there. VP links are concatenated to form a virtual path connection (VPC). A VPC extends between two VPC endpoints. In the case of point-to-multipoint connections, there are more than two VPC endpoints. A VPC endpoint is the point at which the VPIs are originated, translated, or terminated.

When VCs are switched/cross-connected, the VPC supporting the incoming VC links must be terminated first, and a new outgoing VPC must be created. Cell sequence integrity is preserved for cells belonging to the same VPC. Thus cell sequence integrity is preserved for each VC link within a VPC.

3.2.2 ATM Layer Specification

Cell Structure [34]

Cell structure is determined at the UNI and NNI, with a slight difference. The ATM cell consists of a 5-octet header and a 48-octet information field as shown in Figure 3.7. At UNI and NNI, some of the preassigned values of the header are used to identify cells for physical layer functions: idle cell identification, physical layer OAM cell identification, and other physical layer cell identification.

Cell Structure At UNI

UNI assigns four bits of the generic flow control (GFC) field, however, usage has not yet been determined. When the GFC field is not used, the value of this field is 0000. This field will be utilized by the equipment implementing the *controlled transmission* set of procedures.

Twenty-four bits are available for routing: 8 bits for VPI and 16 bits for VCI. Some combinations of VPI and VCI values are reserved for identifying signaling cells, OAM F4 (VP availability and performance monitoring function), and F5 (VC availability and performance monitoring function) flow-related cells [35], and other special-use cells.

The payload type (PT) field is three bits and is utilized to identify PT cells such as user data cells experiencing or not experiencing congestion, OAM F5 segment and end-to-end associated cells, resource management cells, and cells reserved for future functions.

The cell loss priority (CLP) bit is utilized to identify priority for cell discarding, depending on network conditions. Cells whose CLP is set (CLP value is 1 or low-priority) are subject to discard prior to cells whose CLP is not set (CLP value is 0 or high-priority).

The header error control (HEC) field consists of eight bits. HEC covers the entire cell header. The code used for this function is capable of either single bit error correction (default mode), or multiple bit error detection.

Cell Structure At NNI

The major difference in the NNI and UNI cell header structures is the routing field. Since there is no GFC field in the NNI cell header, 28 bits are available for routing; 12 bits of VPI and 16 bits for VCI (same as at UNI). Some preassigned field values are slightly different from those at UNI.

3.2.3 ATM Network Aspects

Application of the VPC

Different types of VPC applications [33] have been identified. They include the user-user application, user-network application, and network-network applica-

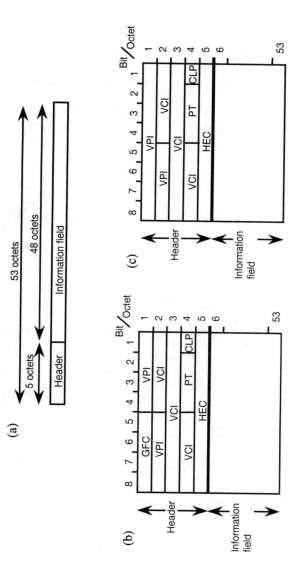

Figure 3.7 Cell structure: (a) cell structure at UNI/NNI, (b) header structure at UNI, and (c) header structure at NNI.

tion illustrated in Figure 3.8. In the user-user application, the VPC extends between T_B or S_B reference points and provides customers with VPCs. Here, T_B and S_B are reference points for the B-ISDN accesses [36,37], as shown in Figure 3.9. Customers can determine the usage of VCCs within the VPC and create their own (leased line) networks with this capability. ATM network elements between the reference points transport all the cells associated with the VPC along the same route in the network. In the user-network application, the VPC extends between the T_B or S_B reference point and a network node, and it is used to aggregate access of VCCs associated with customer equipment (CEQ) to a network element. In these applications, the VPI will be utilized to identify different services such as public switched service, CATV services, and leased line services. In the network-network application, the VPC extends

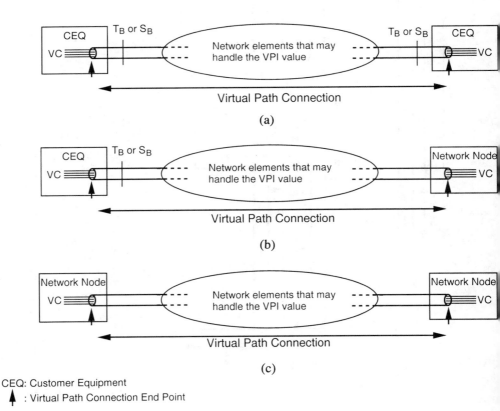

CEQ: Customer Equipment
↑ : Virtual Path Connection End Point

Figure 3.8 Applications of the virtual path connections: (a) user-user, (b) user-network, and (c) network-network.

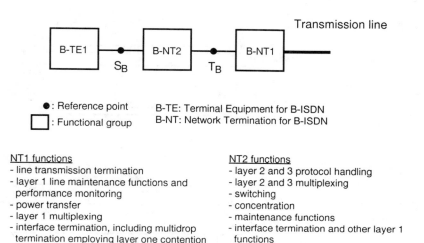

Figure 3.9 B-ISDN reference configuration and NT functions.

between two network nodes. At the network nodes terminating the VPC, the VCs within the VP are switched or cross-connected to VCs within other VPs.

VPs provide logical direct links between virtual path terminators (VPTs), and hence provide the means to manage VCs according to traffic types as shown in Figure 3.10. For example, switching services and leased line services can be separated by differentiating their VPs. This is true for either constant bit rate (CBR) or variable bit rate (VBR) services (see Section 3.4), and for services requiring a high reliability (network OAM information is one example) and other attributes.

In terms of VCCs, similar applications to those mentioned above are possible.

3.2.4 Virtual Path

As mentioned before, simplification of the network architecture coupled with simplification of node processing is the key to developing a cost-effective, flexible network. The VP concept was proposed to fully exploit ATM capabilities and provides the network with a powerful transport mechanism. The concept was first presented in 1987 [38–41] by NTT, and a similar concept, called virtual direct routing (VDR) was provided later [42] by Telecom Australia, whose concept is mostly concerned with ATM cell addressing. The VP concept was proposed to CCITT (now ITU-T) in the January 1988 meeting [41,42].

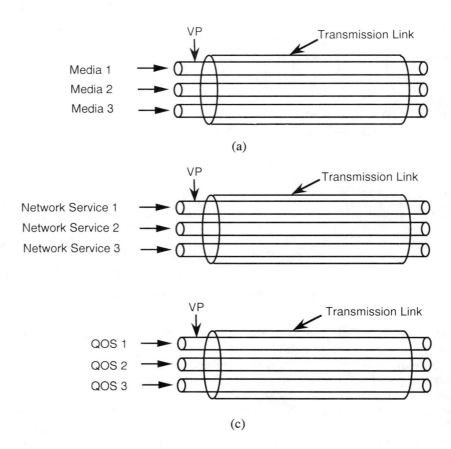

Figure 3.10 Management of different types of traffic by VPs: (a) media management, (b) network service management, and (c) QOS management.

Since its introduction, VP benefits have been subsequently elucidated, and the necessary techniques have been developed allowing VPs to be implemented [8,43–55]. The VP concept was approved as an international standard in CCITT (now ITU-T) in 1990 [56]. Introduction of the VP concept into an ATM network allows management of VCs by grouping them into bundles. Consequently, VCs can be transported, processed, and managed in bundles, which permits many advantages such as reduced node costs and simplification of the transport network architecture. This will be discussed in a comparison with the STM-based network in the next section.

The path is fundamentally a logical concept in the sense that a path is a bundle of circuits, as mentioned above. In STM-based networks, however, only positioned paths carried within a framed interface are possible. In STM networks, specific time slots within the 125-μs TDM frame are assigned to each path. Positioned paths are identified by their time position within a TDM frame, and the capacity of the path is deterministic only. Thus, a path is tightly linked to the physical interface structure for transmission. STM paths, therefore, exhibit a hierarchical structure that reflects the transmission system hierarchy. This linkage produces problems that will become more and more serious as demand increases for a wider variety of services with a wider range of transmission bit rates.

On the other hand, in ATM networks, VPs are identified by the labels (VPI) carried in each cell header. This mechanism enables the separation of logical path aspects from the physical transmission interface structure, so the fully logical realization of path functions becomes possible. The capacities of VPs are either deterministic or statistical. A VP is established between VPTs using the ATM cross-connect/ADM system. The VPs are cross-connected as the ATM cross-connect system performs cell switching according to the VPI and the VP connect table, which is controlled by the VP administration system. The VPI value in the cell header is assigned link by link through the ATM cross-connect system. VP terminating functions include path usage monitoring and control (see Section 3.4). VPs are established or released dynamically, based on long-term service provisioning, short-term demand, or even immediate demand for alternate routing as in the case of network failure. The ADM system is based on the same mechanism as the cross-connect system.

The VP scheme has several advantages over the digital path scheme in STM networks, not only in terms of the node system structure but also in terms of network architecture and management aspects. Explanations of some of these points are provided in Section 3.3 using a comparison to digital paths in STM networks.

3.3 COMPARISON OF ATM AND STM

The introduction of VPs will impact transport network architecture and network management. A comparison of ATM and STM is summarized in Table 3.1. Comparison details and the potential of ATM are described in this section.

3.3.1 Path Networks and Transport Node Architecture

The direct multi/demultiplexing capability of VPs into/out of transmission links eliminates hierarchical path structures and, therefore, eliminates the hierarchical digital cross-connect systems needed by STM. An example of path layer simplification is illustrated in Figure 3.11. The corresponding resultant

Table 3.1
Impact of Virtual Paths on Network Architecture and Management

Items	ATM Network	STM Network
Network architecture	Nonhierarchical virtual path network and resultant simplified network architecture	Hierarchical multilevel path networks
Node architecture	Simplified and cost-effective node architecture with nonhierarchical cross-connect system	Hierarchical multistage digital cross-connect systems
Path accommodation design within transmission facility network	Simple VP accommodation possible with direct path multiplexing capability into transmission links	Optimization procedure is necessary in path accommodation design considering limited multiplexing stages of installed cross-connect systems at each node
Path establishment and control	No time slot management processing for VP establishment. No processing at cross-connect nodes along VPs for path bandwidth alteration	Time slot management in TDM frames is necessary for path establishment. Time slot management processing is necessary for path bandwidth alteration
Service protection	Easier implementation of hitless path rearrangement	Hitless path protection routing is more difficult
Transmission link utilization	Increased transmission link utilization stemming from the direct (nonhierarchical) VP multiplexing capability into transmission link with single-stage ATM cross-connect system. Link utilization increase made possible with dynamic VP management (adaptive VP route/bandwidth control)	Link utilization deterioration due to the required hierarchical multistage multiplexing of paths. Dynamic path management is more difficult

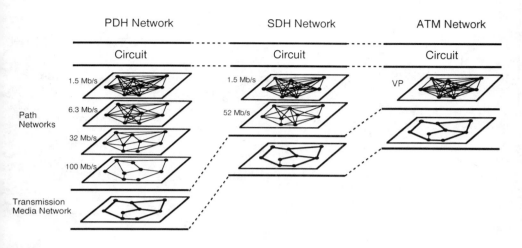

Figure 3.11 Path network simplification.

cross-connect node architecture simplification is shown in Figure 3.12 using Japan's digital hierarchy [57] as an example.

Path layer simplification will reduce node cost and, coupled with the transmission cost reduction made possible by the improvement in link utilization offered by VPs, realize significant network cost reductions. The development of ATM cross-connect systems with large throughput is the key to achieving this. ATM cross-connect system technologies are described in Section 3.5.

3.3.2 Path Accommodation

Paths are accommodated within the transmission facility network using cross-connect/add-drop multiplexer (ADM) systems. Figure 3.13 compares path accommodation in STM and ATM networks. For the STM network, hierarchical path stages exist and digital path accommodation design was performed to optimize the cross-connection stages of paths at nodes in terms of the total network cost [11,12], that is, transmission cost plus cross-connecting cost. This is because not all the nodes in the network will be equipped with all cross-connect system stages. The hierarchical path structure and limited number of cross-connect system stages result in poor link utilization. First, as mentioned previously, hierarchical path multiplexing tends to result in inadequate link utilization. If the utilization efficiency of paths at each stage is 0.7, three multiplexing stages yield the link utilization of 0.7^3; that is, only 34%. Second, the

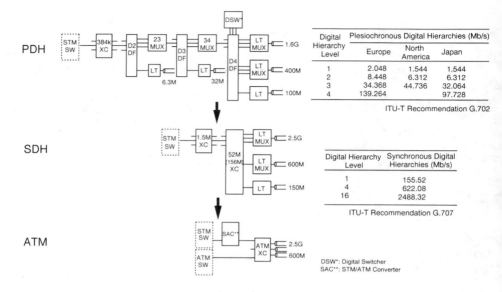

Figure 3.12 Simplification of transport node architecture.

total path length increases since each stage of cross-connect systems is equipped within a limited set of nodes. This increases total link cost.

In ATM networks, on the other hand, there is only one stage of VP and the direct multiplexing of VPs becomes possible with ATM cross-connect systems. This greatly facilitates VP accommodation design, as seen in Figure 3.13, and results in increased link utilization. This is because link utilization is determined by the efficiency of single-stage multiplexing. In ATM, however, sophisticated network resource management is required due to the asynchronous nature of cell multiplexing. Details of ATM network resource management are given in Section 3.4.

As already mentioned, the simplification of the cross-connect node architecture and path accommodation design results in transmission network cost reductions. Figure 3.14 shows a network cost reduction example for a transmission facility network constructed around VP techniques, compared to an SDH network. This comparison is based on a model of the typical regional network in Japan, as shown in Figure 3.13. For the STM network based on SDH, two-stage path networks, 1.5-Mbps and 52-Mbps paths, and corresponding cross-connect systems are required, as shown in the figure [57,58]. In this model, 52-Mbps cross-connect systems are installed in each node; however, 1.5-Mbps cross-connect systems—which are connected to the backbone trunk transmission links—providing connectivity among regional networks, are installed at only

Asynchronous Transfer Mode Network 57

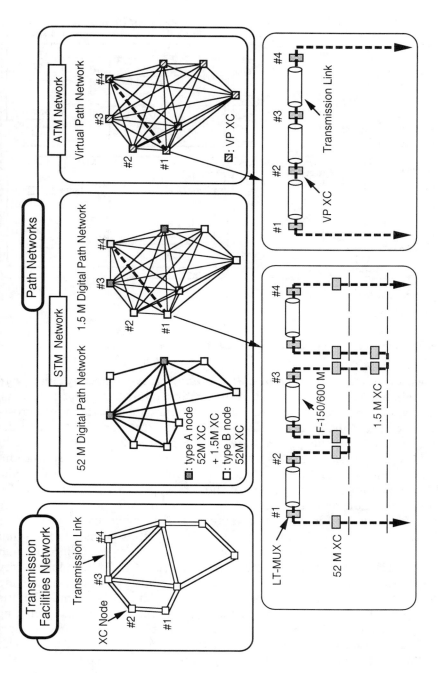

Figure 3.13 Comparison of path accommodation in STM and ATM networks (*After:* [59]).

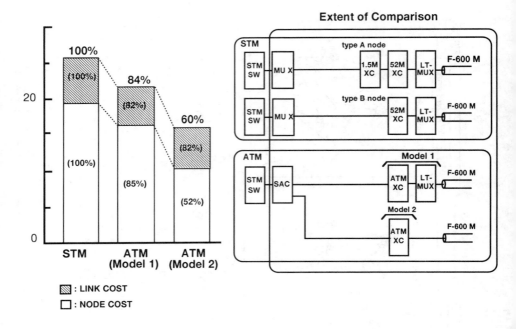

Figure 3.14 Comparison of installation cost in STM and ATM networks.

two nodes, as indicated in the figure. If 1.5-Mbps cross-connect systems were added to all nodes, transmission link utilization could be enhanced. This, however, would increase node cost. In the network model, therefore, cross-connection of 1.5-Mbps paths is only possible when they are fed through the 1.5-Mbps cross-connect or type A nodes (See Figure 3.13). In other words, it is judged less cost-effective to create a direct 52-Mbps path between source and destination nodes of the 1.5-Mbps path to accommodate it than to feed the 1.5-Mbps path from the source node to the 1.5-Mbps cross-connect node for the cross-connection and then feed it to the destination node, even if the 1.5-Mbps path length is consequently increased. This is because the 1.5-Mbps path demand was not large enough to fill direct 52-Mbps paths. On the other hand, the equivalent ATM network consists of a single-stage VP network and single-stage ATM cross-connect systems (see Figure 3.13). This enables every VP to be cross-connected at every ATM cross-connect node. Two models are used for the ATM cross-connect nodes, as shown in Figure 3.14. In model 1, the ATM cross-connect system uses a separate line termination and multiplexing system, while in model 2, the function is included in the ATM cross-connect system. Other assumptions made for the evaluations are that route diversity for 1.5-Mbps paths is employed, the average internode distance is 130 km, the assumed total traffic is 6,750 DS1s, and link utilization for SDH and ATM networks is 1 (the link utilization inefficiency of the STM network caused by hierarchical digital path multiplexing and that of the ATM network stemming from VP resource management, described in Section 3.4, are not considered for simplicity).

The evaluation results (Figure 3.14) show that model 1 attains a cost reduction of 16%. This reduction consists of an 18% reduction in link cost and a 15% cut in node cost. The link cost reduction stems from the total path length reduction [59] made possible by the nonhierarchical cross-connect function of ATM and the increased link utilization mentioned above. The node cost reduction is due to the elimination of hierarchical multistage cross-connect systems. In model 2, the total cost reduction increases to 40%.

3.3.3 Path Bandwidth Control

Figure 3.15 compares path bandwidth control in STM and ATM networks with cross-connect systems. In STM networks, path bandwidth is reserved by allocating time slots in TDM frames at STM cross-connects. The alteration of path capacity and route requires time slot reallocation in the TDM frames at all STM cross-connects along the path. On the other hand, in ATM networks, a VP route is established by setting the routing tables of the ATM cross-connect systems. The path capacity allocation in the transmission facility network is

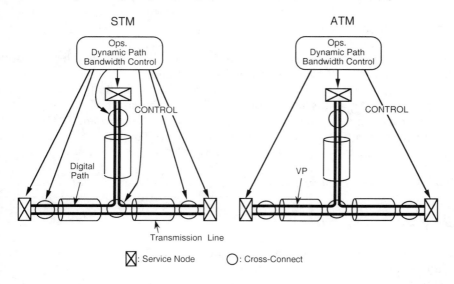

Figure 3.15 Comparison of path bandwidth control in STM and ATM networks (*After:* [52]).

done logically. The VP capacity is reserved in the network through management procedures.

The ATM cross-connect nodes between the VPTs do not engage in VP capacity management. Therefore, alteration of VP capacity requires no processing at cross-connect nodes along the VP, as shown in Figure 3.15. This independence of route and capacity control is an outstanding benefit of VPs and will greatly enhance network capability. An example of this is explained below.

Dynamic Path Network Reconfiguration

Dynamic/adaptive path network reconfiguration will be an important technique to create highly reliable multimedia networks and to enhance the grade of services that the network can provide. The network operating companies can increase responsiveness to changing customers' demands and can provide customer control capability with reduced cost. This customer control application provides freedom for customers to design and manage their own closed networks, which can be created by VPs [54,60].

Path networks can be reconfigured in response to changes in the network state by controlling two parameters: bandwidth and route. Bandwidth control is needed to respond to traffic changes. Route alteration is required for traffic changes and network failures. The VP network offers enhanced adaptability for handling unexpected traffic variations and network failures. This is because of the following:

- VP bandwidth is determined nonhierarchically. This enables optimized or improved path routing and bandwidth control.
- Direct multiplexing of VPs with different bandwidths becomes possible by employing a single kind of ATM cross-connect/ADM.
- No processing is necessary at nodes along the paths between VP terminators for path bandwidth allocation or alteration. The time slot reallocation within TDM frames necessary in STM networks is eliminated.

These are unique characteristics of VPs, and the last point is elaborated below.

VP Bandwidth Control

In ATM networks, switches will be connected by VPs. Offered traffic between switches varies with the time of the day, the time difference within the country, and for unexpected reasons. Moreover, offered traffic fluctuates from minute to minute. If the link capacity is dynamically shared among VPs, the total VP bandwidth required to attain a specific call-blocking probability will be reduced. Therefore, the minimum link capacity required to accommodate the paths is also reduced. An example of the capacity-sharing effect is shown below [44,61–63]. Figure 3.16 depicts the model used in the analysis, where n links are connected by ATM cross-connects in tandem. For simplicity, link capacity, C, the number of virtual paths, N, accommodated by the link, and the average traffic offered to each VP, a erl, are assumed to be constant. Other assumptions adopted to make the analysis easier are as follows:

- The path bandwidth is deterministic (no statistical cell multiplexing effect is considered).
- The call arrival process is a Poisson process, and the bandwidth of every call is 1.

The call-blocking probability corresponding to each path is set to be less than 10^{-3}. The bandwidth of each VP (accommodated in the link) is increased or decreased in steps of "S" according to the number of calls offered to each path. The path bandwidth control algorithm used here is as follows:

- Request a VP bandwidth increase if there is not enough spare capacity to accept a new call arriving to the VP. If the VP bandwidth increase is possible, then increase it and establish a new VC for the call; otherwise, reject the call and keep the VP bandwidth unchanged.
- Decrease the VP bandwidth if it is possible, according to VP utilization.

Different VP statuses are possible for a model with n links in tandem, as shown in Figure 3.16. The call-blocking probability corresponding to each path differs according to the path accommodation status. The worst call-blocking probability occurs for the path (VP 1) that traverses all n links. The call-blocking probability for VP 1 in Figure 3.16(a) is less than that for VP 1 in Figure 3.16(b), where VPs other than VP 1 in Figure 3.16(a) are substituted for the VPs that travel only one link. No formal proof is given here, but this conclusion is based on a wide range of simulations [62,63]. This means that when the blocking probability for VP 1 in Figure 3.16(b) is less than 10^{-3}, the blocking probability for all paths in Figure 3.16(a) is less than 10^{-3} irrespective of the path accommodation conditions. The blocking probability for VP 1 in Figure 3.16(b) can be analytically derived [62,63], and utilizing this, the path bandwidth control effect can be analyzed.

The effect of path bandwidth control was measured by the normalized link capacity (NLC), which is the ratio of required link capacity with control to the capacity required without control. The required capacity is the capacity necessary to satisfy a specific call-blocking probability (in this case, 10^{-3}).

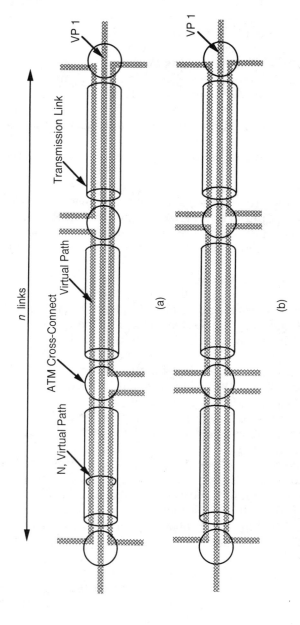

Figure 3.16 Different virtual path statuses and a model used for analysis.

Figure 3.17 shows NLC as a function of the number of links traversed (n), for two different offered traffic values. Here, a was set at 30 and 200 erl, and S took values of 4, 8, and 12. The analysis assumed that the blocking probability was 10^{-3} and $N = 16$. The results plotted in Figure 3.17 indicate that NLC is only slightly dependent on n when the specific blocking probability is 10^{-3}. Analysis details are found in [62,63].

The increase in processing load created by path bandwidth control was also analyzed. The normalized processing load (NPL) is defined as the ratio of the frequency of path bandwidth change requests to that of call setup requests. NPL varies with S, however, it is insensitive to n. NPL values for a of 16 erl and 200 erl were 0.126 and 0.124, respectively, when S was 8 and n was 8. These results show that link capacity utilization can be increased with an insignificant increase in processing load under the practical path condition that more than one link is traversed.

Remember that, as described before, VP bandwidth control requires no processing at cross-connect nodes along the paths, and no synchronization processing between the sender node and destination node is required either. This reduces the time required to control path bandwidth and, therefore, increases network responsivity against abrupt traffic changes.

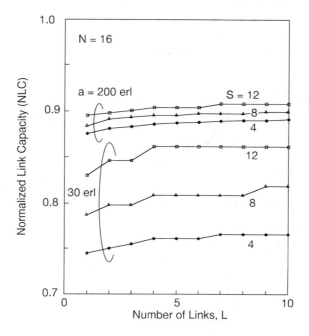

Figure 3.17 NLC as a function of number of links traversed.

Fallback on VP Application

The technique described above can be applied to the protection routing of VPs [50], as shown in Figure 3.18. In this application, protection routes can be pre-established without reserving capacity [72,73]. That is, the VPI numbers associated with each link along the protection path routes are reserved and preassigned in the path connect tables at cross-connect nodes along the paths. A certain link capacity should be reserved for protection path use, but the reserved capacity can be shared among many possible protection paths accommodated by the link. When a network failure is detected, the restoration path bandwidths are reserved logically through a centralized management process, and cells are switched to the protection path route by providing new VPI values specific to the protection path. Of the transmission facilities, only the originating VP terminator (in case of path route switching at the originating end point of path; this is schematically illustrated in Figure 3.18) or the ATM cross-connect (in case of path route switching at failed link end) is involved in the process of routing alteration. No routing table renewal or other processing at cross-connect nodes along the protection path is required, and this enables rapid restoration.

In this process, the fallbacks on path routes are predetermined. However, these routes can also be determined on a real-time basis using distributed processing, as is employed by self-healing network techniques.

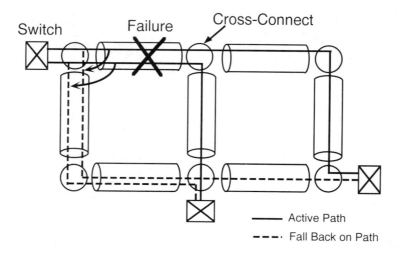

Figure 3.18 Fallback on path application (*After:* [50]).

VP Self-Healing

By exploiting the logical realization of VPs, self-healing performance with distributed control can be enhanced. The primary advantage of self-healing is that it permits more rapid restoration than is possible with existing path restoration schemes that use centralized control mechanisms. Existing self-healing techniques developed for STM networks use digital paths [64–70]. Existing self-healing algorithms require at least one round-trip exchange of restoration messages between sender and chooser nodes (restoration pair nodes) to determine the restoration path route and to assign its capacity, as shown in Figure 3.19. Here, the sender node and chooser node are the end nodes of the restoration segment of a failed path. On the other hand, in the self-healing algorithm newly developed for VPs [71,72], restoration path establishment is completed with the transmission of a restoration message in just one direction. This is because one-way message transfer, from the sender node to chooser node, can determine candidates of multiple restoration path routes. The restoration process is briefly explained as follows. When VPs are cut by failures, the sender node (the node

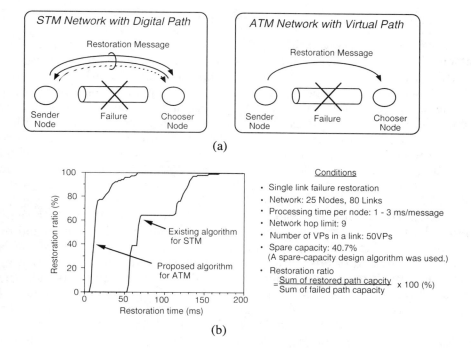

Figure 3.19 Comparison of self-healing techniques: (a) self-healing schemes in STM and ATM networks and (b) restoration characteristics (*After:* [71]).

immediately downstream of the failed link) checks the spare capacity of each link connected to it and establishes possible restoration routes, but the bandwidths on these routes are not assigned (zero bandwidths). The sender node connects the potential VPs to the downstream part of the failed VPs (unfailed parts of VPs). The sender node then broadcasts a restoration message through all the links connected to it. The restoration message contains information such as the following [71]:

1. ID number of the failure as defined by the sender node;
2. Number of failed VPs;
3. Total failed VP capacity;
4. Hop count;
5. IDs of links traversed by the alternate path;
6. Reserved capacity of the alternate path in each link;
7. Restoration priority value as assigned by sender node (needed for multiple failure restoration).

At nodes that received the restoration message, the spare capacity of each attached link is checked and the restoration VPs are extended to the following links, and thus restoration VPs are developed from downstream to upstream of the failed VPs. These processes are done node by node until the message reaches the chooser node (immediately upstream of the failed link). The chooser node calculates the available capacity of the alternate route and the bandwidths already allocated to other restoration VPs on the route. The chooser node chooses which failed VPs will be restored based on a predetermined selection rule, and bandwidths are allocated to the restoration VPs. The selected VPs on the restoration route are connected to the corresponding upstream part of the failed VPs and thus the VPs are restored. The cross-connect nodes along the restoration VP do not need to be informed of the new restoration VP capacity before the failure is restored. Consequently, restoration can be completed by transmitting restoration messages in just one direction. After the failure is completely restored or after a timeout, the chooser node sends a terminating message through the adopted routes so that the spare capacity is reduced by the bandwidth of the adopted restoration VPs. This technique exploits the logical aspects of a VP: the independence of route and capacity establishment. This is not possible in STM networks. This simple restoration process can reduce the restoration time required. Figure 3.19 compares the restoration performance realized by the algorithm for VPs with the comparable technique for STM networks. Details of the algorithm are provided in [71].

3.3.4 Subscriber Networking

So far we have discussed backbone networks. The VPs are also effective in creating multiservice networks for subscriber access. The point is that different services with different service nodes such as public switched services, leased line services, and center-to-end video distribution services can be provided through a single subscriber network in an integrated fashion. This is made possible by virtue of the logical service separation capability of VPs on common physical links.

Several types of physical network configurations have been proposed for future access transport networks. A promising configuration combines a transfer network with loop networks in which a passive optical network (PON)/passive double star (PDS) loop [74–76] or single star architecture is adopted. Figure 3.20 depicts a possible access transport network. A VP directly connects customer premises' equipment to desired service nodes. It is very easy to provide the desired combination of services for each customer by VP control; in Figure 3.20, three types of VPs—for video distribution, switched service, and ATM leased line service—are provided simultaneously through a single user-network interface. This enhances network utilization and flexibility for

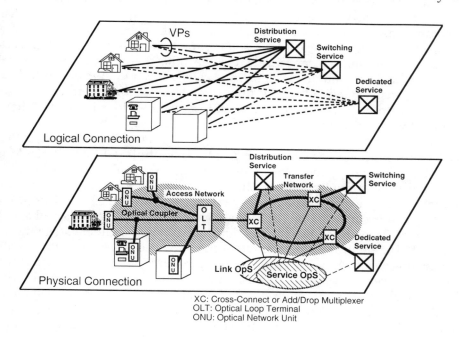

Figure 3.20 Access transport network with VPs.

unexpected service demands. Thus, effective subscriber networks can be developed by capitalizing on VP capabilities.

3.4 NETWORK RESOURCE MANAGEMENT

The path is fundamentally a logical concept in the sense that a path is a bundle of circuits. In PDH and SDH networks, a path is tightly linked to the physical interface structure for transmission and exhibits a hierarchical structure. The VP strategy, on the other hand, enables the fully logical realization of path functions. In ATM networks, as explained in Section 3.1.4, VP capacity is nonhierarchical, and direct multiplexing of paths into transmission lines becomes possible by using a single cross-connect system. Thus in ATM, the channel and path are logically realized, which enables more effective cell transfer supporting various speed (bandwidth) and burstiness than STM. ATM is particularly superior in the development of multimedia communication networks. However, a new development in network resource management is necessary to design and to operate ATM networks. This requirement stems from the inherent asynchronous nature of cell transfer.

In an STM network, resource allocation or accommodation design for circuits and paths is straightforward and requires no special techniques. This is because the bandwidth for each circuit and path is deterministic and hierarchical. Allocation is physical by assigning the time slots within 125-μs TDM frames to the required bandwidth. On the other hand, in ATM networks, VP/VC bandwidth is nonhierarchical, and cells on each VC/VP are multiplexed asynchronously. Bandwidth allocation, therefore, is done logically (no time slot or cell slot is physically allocated or reserved). Various issues peculiar to ATM networks arise from this.

One example of this is depicted in Figure 3.21. In STM networks, if there are 24 64-Kbps incoming channels/paths to be multiplexed into a 1.536 Mbps (= 64 Kbps × 24) outgoing transmission line, no information loss occurs (here jitter and wander due to transmission line systems are assumed to be 0, and 1-octet memory exists for each incoming port). However, in ATM networks, cell loss can occur when multiplexing constant bit rate channels/paths, even if the total capacity of incoming channels/paths is less than outgoing link capacity, if the multiplexing cell buffer memory size is less than the number of incoming channels/paths. This occurs, for example, if a cell from each incoming port arrives at the same time as shown in Figure 3.21; the probability of this is small. In this example, the necessary cell buffer size to completely eliminate cell loss is 24. When the outgoing link capacity is large, a quite large cell buffer is required to eliminate cell loss. Available buffer sizes are, unfortunately, limited even with state-of-the-art hardware. Thus the minimum required buffer size must be determined according to the cell loss specification

Asynchronous Transfer Mode Network 69

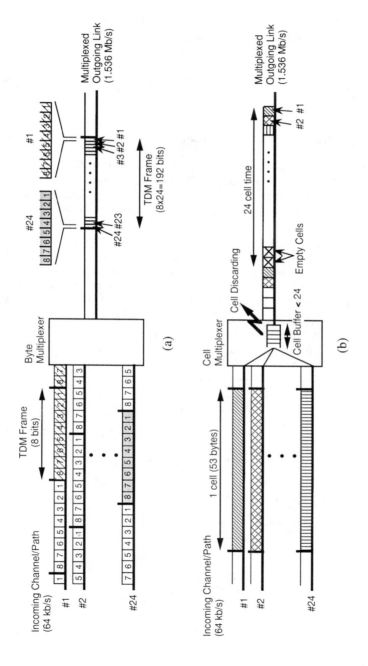

Figure 3.21 Multiplexing examples in (a) STM and (b) ATM.

at multiplexing, considering incoming and outgoing channels/paths capacity and the target outgoing link utilization even for constant bit rate channels/paths. The difficulty of designing the necessary cell buffer size is greatly worsened when the incoming information traffic is bursty (VBR). ATM network resource management techniques including the above problem are described in this chapter.

3.4.1 Network Resource Management Principle

The ATM network resource management principle is depicted in Figure 3.22. The basic idea is the same as that in STM networks. The resource management level can be divided into three levels. In the first level, VC traffic demands in the network and a specific call-blocking probability (VC) determine switching facility demands and the necessary number of circuit demands between switching system units; these requirements determine the circuit network. VP bandwidths between switching units are determined so that the necessary number of VCs can be accommodated and the necessary VC quality of services (QOSs) (cell loss rate, cell transfer delay, etc.) can be satisfied. The created VP network is accommodated within a transmission facility network so that VP QOSs and VP reliability requirements (VP restoration) can be satisfied.

In established connection-oriented ATM networks, VC/VP traffic management [77] will be divided into two phases (see Figure 3.23). In the connection

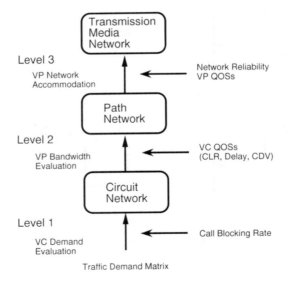

Figure 3.22 ATM network resource management.

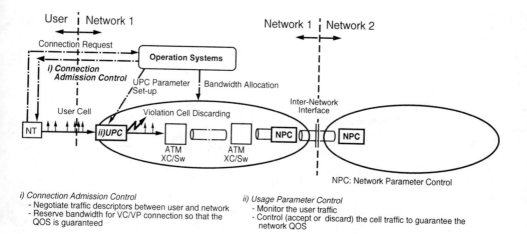

Figure 3.23 ATM network resource management principle.

setup phase, the connection admission control mechanism decides whether the new VC/VP connection should be accepted or not based on the VC/VP bandwidth, the QOS requirements, and the network's available bandwidth. In this process, the bandwidth requirement of the VC/VP is indicated in the form of traffic source descriptors at T_B reference point or can be implicitly declared by the service type, which may include traffic descriptors and QOS requirements. Upon acceptance, the network reserves the necessary bandwidth along the end-to-end VC/VP connection so that the QOS to be provided by the network is satisfied. In the information transfer phase, traffic generated by the source is monitored and controlled (cells sent to the network are accepted or rejected according to violation of the negotiated traffic descriptor values) by the network usage parameter control (UPC) function (when this function is performed at internetwork interfaces, it is called network parameter control (NPC)), so that the QOSs of other connections are not degraded. In this process, cell priority control can be performed. The important point is that the value of the UPC parameters and the network resource allocation based on the parameters can be determined by the network on the basis of the network operator's policy. For example, for VBR traffic, the network may allocate the network resources and control the user traffic based only on the peak rate, thus minimizing the necessary processing at the cost of lowering the network resource utilization efficiency.

Source Traffic Descriptor

A source traffic descriptor [77] is the set of traffic parameters that will be used during the connection setup phase to represent the traffic characteristics of VP/VC connection. The parameters should be as follows:

- Easy for the user to understand; in other words, conformance should be possible;
- Effective in network resource management;
- Enforceable by the UPC and NPC.

In general, the more precisely the user (terminal) defines the traffic characteristics, the more effectively the network resources will be managed. In other words, for the bursty traffic, the more information given to the network on the traffic characteristics, the less resources the network can allocate to the connection to attain specific QOSs (and then the transmission cost can be reduced). The most important traffic descriptor is the peak cell rate. Others include average cell rate, peak duration, and burstiness; some of these parameters are dependent on each other. To specify average cell rate, two parameters are required; the average rate of 0.5 has little meaning if you do not specify the averaging time—if you send traffic at intensity 1 for the first year, then at 0 the next year, the two year average is 0.5. One possible set of traffic descriptors and the corresponding UPC scheme [82] is shown in Figure 3.24. For the first ITU-T Recommendation [77], only the peak cell rate is defined.

Reference Configuration and Equivalent Terminal

The peak cell rate is defined as the inverse of the minimum interarrival time between ATM cell emission requests at the physical layer service access point (SAP) for an equivalent terminal representing the VPC/VCC, as shown in

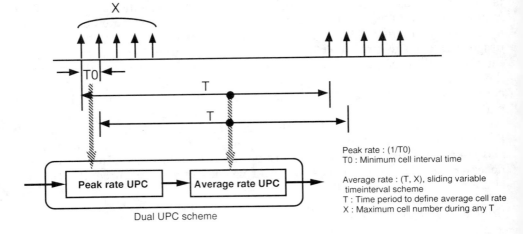

Figure 3.24 Usage parameter control.

Figure 3.25. The point to be noted is that cell multiplexing will alter the traffic characteristics of ATM connections by introducing cell-delay variation (CDV), as shown in Figure 3.26. When cells from two or more ATM connections are multiplexed, or physical layer overhead or OAM cells are inserted, cells of a given ATM connection may be delayed. Therefore, UPC/NPC functions and network resource allocation must consider this CDV effect. The maximum allowable value of CDV between the ATM connection endpoint and T_B, between T_B and an internetwork interface, and between internetwork interfaces must be standardized.

Traffic Shaping

Traffic shaping [78–81] is a mechanism that alters the traffic characteristics of a cell stream on a VCC/VPC. Of course, traffic shaping must maintain the cell sequence integrity of each connection. Examples of traffic shaping include reducing CDV by suitably spacing cells. This makes UPC easier and reduces the bandwidth that must be allocated to the connection. Traffic shaping may be used to reduce the peak cell rate. Traffic shaping may be done by users or by network operators. The user may apply it to ensure that the cells generated by the source or at the UNI conform to the negotiated traffic contract or to reduce the peak cell rate for cost reasons. The drawback of shaping is the

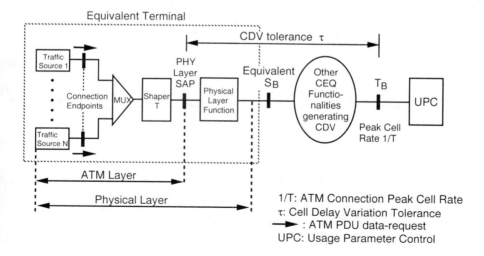

Figure 3.25 Reference configuration and equivalent terminal for the definition of the peak cell rate of an ATM connection.

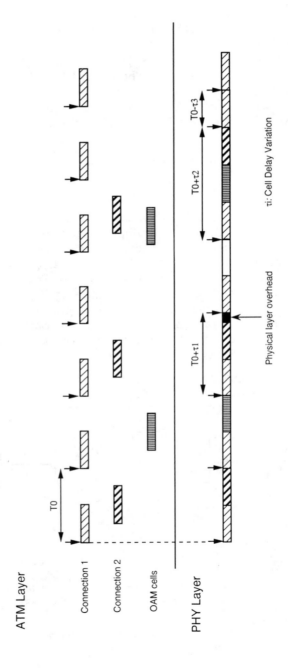

Figure 3.26 Origins of cell-delay variation.

increase in cell delay. If it is done in the network, the increased delay has to be considered in the end-to-end QOS design.

Usage Parameter Control

UPC/NPC should accommodate the effects of CDV. Network resource allocation will be done taking account of the worst case traffic passing through UPC/NPC in order to avoid impairments to other ATM connections. This worst case traffic depends on the specific implementation of the UPC/NPC, which can be operator-dependent. Therefore, tradeoffs exist between UPC/NPC complexity, worst case traffic, and optimization of network resource allocation. The choice is at the discretion of the network operator. In Table 3.2, different UPC algorithms are compared in terms of performance and hardware complexity [82–84]. In this example, two traffic descriptors, peak cell rate ($1/T_0$) and average rate (X, T), are assumed to be controlled. To monitor and control the three parameters (T, X, T_0), a dual UPC algorithm [82] (Figure 3.24) is used; first, peak rate, and second, average rate. Here, only average rate UPC is discussed.

In the Credit Window method, the number of cells during time T is counted with a deterministic phase. This algorithm is advantageous because of its small hardware requirement; however, it has the shortcoming of mispolicing, as shown in Table 3.2, the method can accept a violating cell traffic pattern (more than X cells in a T period). Therefore, in resource management, bandwidth reservation for the connection should be done based on the possible mispoliced worst pattern; $2X$ burst cells in a $2T$ period. This cell pattern is more bursty than X burst cells in T period and so requires more bandwidth, nearly twice as much [84]. The final bandwidth depends on various parameters such as T, X, multiplexing buffer size, and specific cell loss rate at multiplexing. To allow more precise UPC, a variation has been proposed [84]. The method uses two Credit Window circuits with the window phases being offset by $T/2$. The method significantly reduces mispolicing while its hardware implementation remains simple; only a register is needed to store the cell counts in each $T/2$ period.

The second algorithm uses a Sliding Window for cell counting; cells are counted during any period of T using the FIFO (shift register) function. This achieves deterministic UPC; however, a large amount of hardware is required when T is large.

The third is known as the Leaky Bucket (LB) method [20,85]. This algorithm monitors the traffic mean rate by leak rate, R, of a bucket and traffic burstiness by bucket size, B. Figure 3.27 shows the flowchart of the algorithm, where i is an increment value. The LB method uses simple hardware [84] and, as a result, many studies have been done. However, it is known that this method allows *leak patterns* for each set of (T, X) values. The worst leak pattern from the bandwidth allocation point of view has not yet been identified [86]. In other

Table 3.2
Comparison of Different UPC Schemes

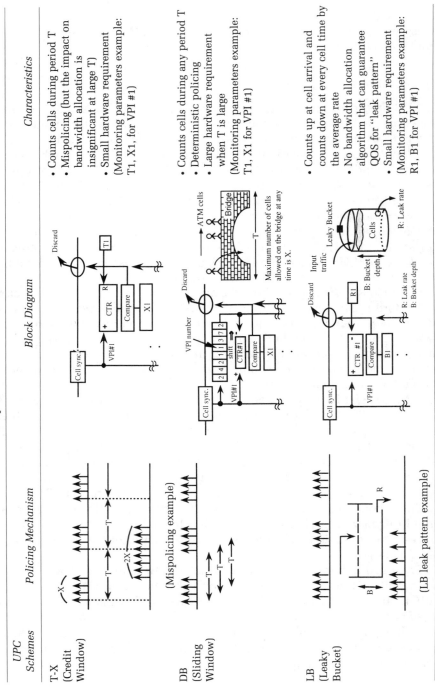

Source: [84]. Reprinted with permission.

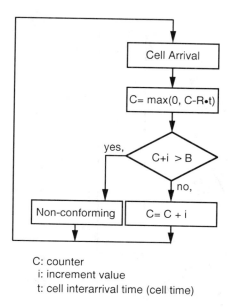

C: counter
i: increment value
t: cell interarrival time (cell time)

Figure 3.27 Leaky Bucket algorithm.

words, an effective and conservative bandwidth allocation procedure has to be created to ensure the network QOS and maximize efficiency. This is explained below. For the UPC of (T, X), R, B, and i are set to X/T, $X-(X-1) \times T_0 \times X/T$, and 1, respectively. Examples of the cell patterns conforming to these LB parameters are shown in Figure 3.28 and Figure 3.29. The pattern in Figure 3.28 (called the basic pattern, hereafter) is identical to the worst case traffic of the Sliding Window algorithm and was thought to be the worst pattern; however, different patterns have been experimentally confirmed to require more bandwidth than the basic pattern. Actually, the LB algorithm passes more than X cells in time period T for the patterns called leak patterns [86]. The pattern in Figure 3.29 is one of the leak patterns and is constructed by combining the basic pattern and a *child* pattern. The child pattern can be characterized by the (Tc, Xc) values of $Tc = T \times N/M$ and $Xc = X \times N/M$, where N and M are integers and $N < M$, as shown in Figure 3.30. Figure 3.31 shows the survivor function of queue length when 99 VP/VC connections, source traffic descriptors (T_0, T, X) of (10, 1000, 10), are multiplexed. The survivor function for multiplexing the basic and leak patterns is shown. It is shown that the basic pattern is not the worst case pattern, so bandwidth allocation based on the basic pattern cannot always guarantee the network QOS. Unfortunately, it has not yet been determined which leak pattern is the worst pattern.

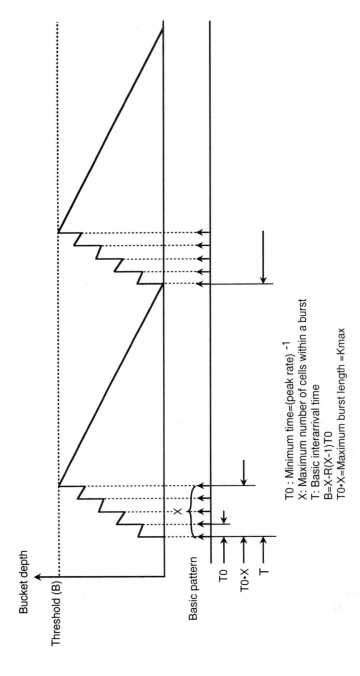

Figure 3.28 The basic or *the worst pattern* permitted by LB (*After:* [105]).

Asynchronous Transfer Mode Network 79

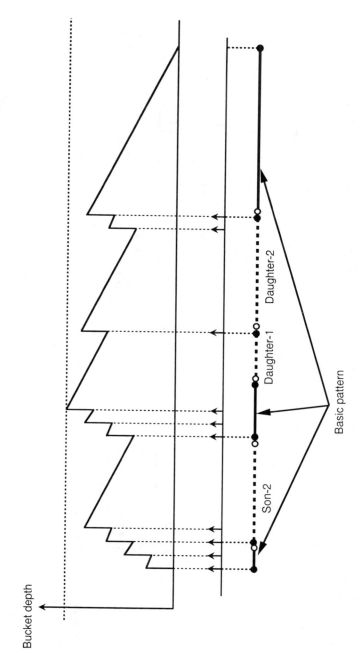

Figure 3.29 Example of a combined pattern of the leaky basic and some child patterns (*After:* [105]).

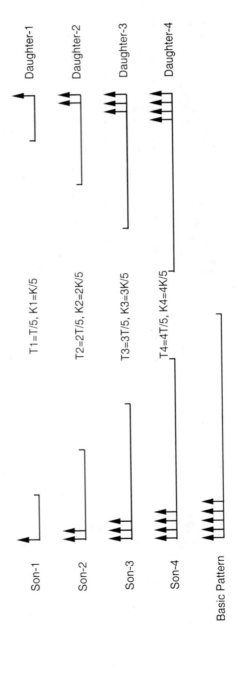

Figure 3.30 Some variations of the basic pattern (child patterns) (*After:* [105]).

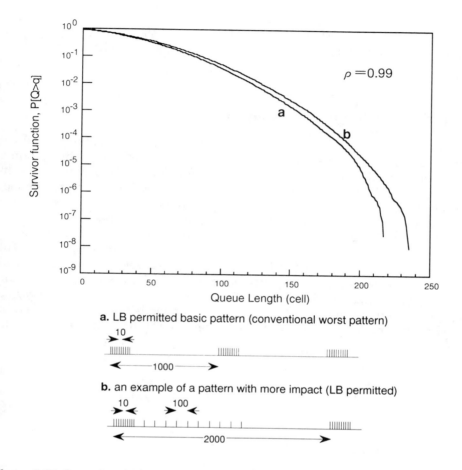

Figure 3.31 Queue length simulation results of (a) *the worst pattern* and (b) the newly discovered LB permitted pattern with more impact (*After:* [86]).

As explained before, CDV will be incurred when two or more ATM connections are multiplexed before the UPC or physical layer overhead or OAM cells are inserted. An appropriate parameter margin is needed to avoid overpolicing at UPC if CDV is present because even if the user adheres to the negotiated traffic pattern, jitter may make traffic appear to violate the UPC. One effective UPC method is the parameter conversion method [84]. In this method, the UPC parameters are obtained by conversion from source traffic descriptors and the amount of CDV. The UPC parameters are based on the worst clumping pattern examples as explained in [84]. A large CDV has a large, negative impact on network efficiency, especially with high-speed CBR traffic. In other words, low-speed CBR traffic is CDV-tolerant. Therefore, if the CDV value is large, a traffic shaper is effective in controlling the CDV value especially for high-speed CBR traffic.

3.4.2 Network Resource Sharing Principle

To utilize network resources effectively, various levels of network resource-sharing mechanisms have been implemented. Figure 3.32 illustrates the time scales and the corresponding resource-sharing levels utilized with regard to the time scales. These different levels can be utilized in combination. Connection level sharing is employed in existing networks with switching systems and is a basic function in public telecommunications networks.

Burst-level sharing [77,87–89] has been utilized in LANs with media access control (MAC) functions. In ATM networks, the fast resource management functions can be more easily implemented than in STM networks since logical network resource allocation is possible as described in Section 3.3.3. This management utilizes OAM cells and node-by-node distributed processing instead of existing signaling procedures to minimize the processing delay. In the admission control, in terms of burst-level control, only the peak cell rate

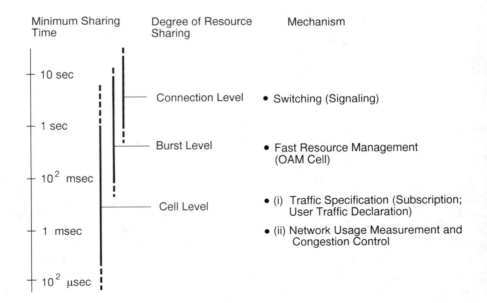

Figure 3.32 Network resource-sharing principles.

of the burst will be utilized in order to simplify admission control and to minimize the incurred delay.

Cell-level sharing produces statistical cell multiplexing gain, which can reduce the resources allocated to the VCC/VPC. Different cell-level sharing schemes are possible. One is to utilize other source traffic descriptors in addition to the peak cell rate to capture the statistical cell traffic characteristics. The network allocates resources to the connection based on an appropriate bandwidth allocation algorithm. The UPC guarantees user traffic to conform to the declared traffic descriptor values, and thus guaranteed QOS will be achieved while also achieving statistical gain (see Figure 3.33). In this scheme, no network congestion due to the established connections occur, except for abnormalities. Another approach is to employ the real-time network traffic measurement scheme [90]. For example, the user declares only the peak traffic descriptors and the network performs CAC based on both the declared values and the actual usage of the network resources as measured on already established connections. In this scheme, network congestion can occur, so a network congestion control mechanism is required. The QOSs that the network provides are objective values and not the guaranteed ones.

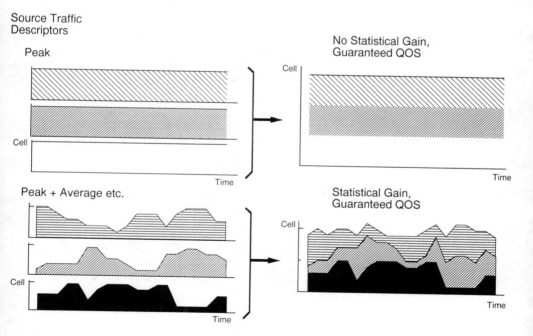

Figure 3.33 Cell-level statistical multiplexing.

Figure 3.34 illustrates these network resource-sharing schemes and the resource reservation principles. Schemes one and two give guaranteed QOSs, while scheme three yields only objective values. Descriptions of each scheme are summarized in Table 3.3.

3.4.3 CBR Path Accommodation Design [91]

Applications of VPs and Bandwidth Attributes

The three main applications of VPs (see Figure 3.8) are for network-network connections, user-user connections, and user-network connections. Network-network VPs are mostly used to link switching units as shown in Figure 3.35. The required bandwidth of these VPs is usually determined from long-term service provisioning needs or, more unlikely, on short-term demand. VPs are cross-connected at ATM cross-connect nodes where cells are switched according to their VPI. User-user VPs typically support leased line services [92,93] and can be used to create private networks. User-network VPs will be

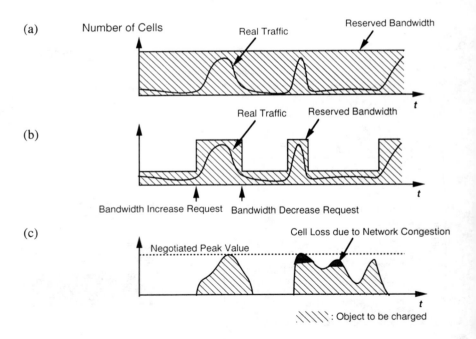

Figure 3.34 Network resource reservation for (a) scheme I (peak rate STD case), (b) scheme II, and (c) scheme III.

Table 3.3
Different Network Resource Management Schemes

	Scheme I	Scheme II	Scheme III
Network resource	Bandwidth reservation for connection	Fast bandwidth reservation for connection (burst)	Nondeterministic bandwidth reservation
Traffic descriptors	Peak, average, etc. (standardized values)	Peak	Peak
Connection admission control	User-declared source traffic descriptor values	User-declared source traffic descriptor values	User-declared STD values + Network traffic measurement
QOS of connection	Guaranteed value	Guaranteed value	Objective value
Cell-level congestion control	Not necessary except for abnormal state	Not necessary except for abnormal state	Necessary
Charging	Reserved bandwidth and duration	Peak rate and duration	Number of delivered cells
Statistical multiplexing effects	None or connection level (peak STD) Cell level (with average STD)	Connection or burst level	All levels (connection/burst/cell levels)
Service	ATM multimedia leased line services	ATM multimedia leased line services (fast reservation)	ATM PVC services

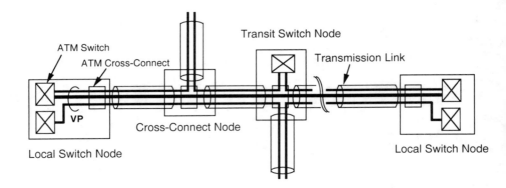

Figure 3.35 Network-network application of VPs. (*Source:* [91]. Reprinted with permission.)

utilized for service access. Services are logically separated on the VPs supported by the same common physical link. This provides the means to construct cost-effective subscriber networks [94,95], as explained in Section 3.3.4.

There are two kinds of VPs in terms of bandwidth [50,77]. The first are called here CBR paths, and their bandwidths are defined by the peak cell rate. The others are the VBR paths, the bandwidths of which are defined by a set of parameters that include the peak cell rate, the average cell rate, and so on. Statistical multiplexing is possible with VBR VPs [50], and hence link utilization can be increased.

This section focuses on CBR VP management techniques. This is because at the outset of ATM introduction, only CBR VPs will be utilized, and the network-network VPs will mostly use CBR VPs even in the mature ATM environment. All STM network paths are CBR paths, and VPs will replace them. Therefore, the development of a VP accommodation design for CBR VPs is of prime importance.

Issues of VP Accommodation Design

The primary attributes of a VP are its route, bandwidth, and QOS. The cell loss rate and cell delay (delay variations) are the major QOS factors. Cell loss is caused by buffer overflow at ATM nodes and bit errors in cell headers, while cell delay is caused by queuing at ATM node buffers (variable delay), processing time, and propagation delay (fixed delay). In terms of cell delay, the CDV stemming from queuing is the most important issue when designing ATM networks. In most VP applications, each end-to-end VP connection must have its own QOS guaranteed. That is, specific values of cell loss rate (typically, 10^{-8} or less), end-to-end cell delay, and CDV will have to be met. These specific values will be standardized by ITU-T.

In order to develop a VP accommodation design algorithm that guarantees the QOS of all VPs, there are several important points that must be resolved, as is schematically illustrated in Figure 3.36. At the VP source node, a mixture of VPs with a wide variety of path bandwidths are multiplexed (cross-connected). The adding or dropping of cells at each cross-connect node causes random variations in the cell-delay of individual VPs. This means that the equivalent bandwidth of each CBR VP increases with the number of nodes traversed from the bandwidth at the VP source endpoint. Thus, CDV at the VP destination node is different for each VP depending on the history of the VP: how many nodes were traversed and what was the traffic condition at each cross-connect node traversed.

Link utilization can be optimized link-by-link by incorporating all the above-mentioned traffic conditions at each cross-connecting point for all the VPs to be multiplexed on a link. This is, however, significantly difficult to do for large networks. This section introduces an approximation method that is very effective because of its simplicity and insignificant drop in link utilization.

Modeling of CBR Cell Traffic

The CBR VP is defined by its minimum cell interval T_0. In the following, T_0 is normalized by the cell transfer time. The multiplexing characteristics of

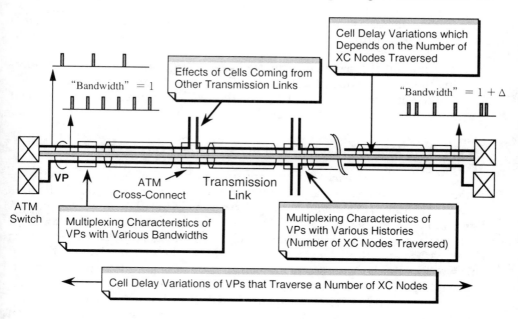

Figure 3.36 Technical points for VP accommodation design (*After:* [91]).

CBR-VPs/VCs without any CDV are evaluated as the solution of N D/D/1 [96,97] or $\Sigma_i\ D_i/D/1$ [98,99] queuing models. For example, Eckberg [96] and Bhargava and others [97] gave an exact solution to the superposition of randomly phased CBR streams that have the same T_0. Robert [98] and Nakagawa [99] provided an approximation of the multiplexing characteristics of multirate CBR traffic. These analytical results, however, are valid only if all the CBR VP/VC cells do not suffer any delay variation. Thus, it is impossible to use them as the basis for bandwidth allocation techniques for regular CBR VPs/VCs that suffer jitter. In general, cell-delay variation causes cell clumping in CBR traffic. This clumping makes the CBR traffic bursty, and analyses that cannot consider CDV underestimate the bandwidth required. The method that is discussed here utilizes the M/D/1 model for the accommodation design of jittered and nonjittered CBR VPs/VCs. It is shown from an exact analysis [96,97] that in the case of a single T_0 value, the smaller $1/T_0$ is (smaller VP bandwidth), the larger the cell loss probability (or $P[Q > q]$; survivor function of queue length) becomes, when the traffic load (link utilization) is constant. In this analysis, Q is the queue length and q is the multiplexing buffer size. Although not theoretically proved, it was numerically shown [96,97] that $P[Q > q]$ converges to the result given by the M/D/1 model with the same traffic load when $1/T0$ approaches zero. This also holds true for the case of multiple T_0 values [98,99]. In other words, the upper bound of cell loss probability $P[Q > q]$ for CBR VP multiplexing is that obtained from the M/D/1 model at the same traffic load.

To use the M/D/1 model for VP accommodation design, the major points that should be clarified and discussed hereafter are as follows.

- The amount of capacity loss caused by the M/D/1 approximation, considering the multiplexing of nonjittered CBR traffic at the first node;
- Conservativeness and accuracy of the M/D/1 approximation at transit nodes that accommodate jittered VPs.

Multiplexing Characteristics of VPs Experiencing CDV

The CDV occurring in each CBR VP increases with each ATM cross-connection. Figure 3.37 shows the effect of buffer size on the maximum traffic load as indicated by the M/D/1 model (the approximation of the CDV effect is explained later) and by strict analysis [96,97] for CBR VPs with a single T_0. The maximum traffic load was determined as that which satisfied the required cell loss rates of 10^{-9}, 10^{-10}, and 10^{-11}. The maximum load was derived from the calculated survivor function for an infinite buffer size instead of a finite buffer. However, it is well known that this approximation yields a conservative cell loss rate.

From the viewpoint of the queue length distribution (or required buffer size to attain specific cell loss rate), it is shown that the M/D/1 model

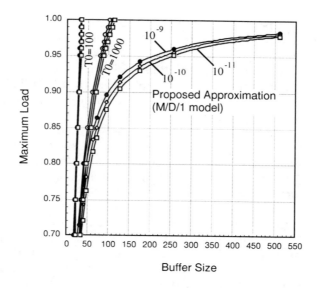

Figure 3.37 Example of accommodation design of CBR VP/VC. (*Source:* [91]. Reprinted with permission.)

approximation yields rather large overestimations [96], especially when T_0 is small. However, from the viewpoint of link utilization, it can be concluded that the $M/D/1$ model yields a fairly good approximation when the buffer size exceeds, say, about 100. That is, the exact analysis achieves 100% of the link utilization when $T_0 = 1,000$ and the buffer size is more than 120, while the $M/D/1$ approximation achieves more than 90% of the link utilization when the cell loss rate is 10^{-9} and the buffer size is greater than 110 cells. Note that from the hardware implementation viewpoint, a 110-cell buffer is easy to attain [54,93]. Thus, the drop in traffic load compared to the results for VP multiplexing in the absence of CDV can be very small, even considering the practicality of hardware implementation.

It is prohibitively difficult and impractical, if not impossible, to strictly analyze the multiplexing characteristics of an actual VP experiencing CDV. The reasons are summarized below:

- The value of cell delay variation differs among the VPs.
- The clumping effect due to CDV varies with the peak cell rate of the VP, even if the CDV value is the same.
- The value of CDV is determined by the number of nodes traversed and the traffic conditions at each node, which include other cross-connecting (multiplexing) VP conditions and the link utilization.

Therefore, it is proposed [91] to approximate the jittered cell traffic of each CBR VP as Poisson traffic where the jitter is caused by the cross-connection (multiplexing/demultiplexing) of CBR VPs on a FIFO basis. This approximation is reasonable because extensive computer simulations have shown that it accurately yields the upper bound of cell loss probability [100,101]. This approximation is also seen as intuitively correct since the periodic cell stream of each VP is modified by the multiplexing/demultiplexing of other VP cells (VP cross-connection) so as to achieve totally random traffic when the number of cross-connect nodes traversed approaches infinity. It is also reasonable to assume that the burstiness of the VP never exceeds that of random traffic provided the cross-connected VPs are CBR VPs. The result of a computer simulation performed to demonstrate this is shown in Figure 3.38.

Figure 3.38 shows the relation between the squared coefficient of variation of cell interval and the number of (cross-connect) nodes traversed as determined by computer simulations for three T_0 and ρ (link utilization factor) values. The simulations assumed that the CBR VP traffic was multiplexed with other Poisson traffic at each node. The simulations were run for tens of thousands of CBR cells. To assume Poisson traffic is fairly severe because a truly realistic situation, jittered CBR VPs produced by multiplexing CBR VPs on a FIFO basis, is less bursty than Poisson traffic, as mentioned before. Therefore, the Poisson traffic

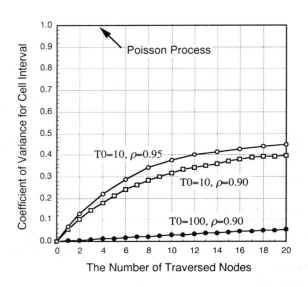

Figure 3.38 Coefficient of variance for the output cell interval. (*Source:* [91]. Reprinted with permission.)

assumption yields larger cell jitter than is realistic. In other words, the result will be the worst case evaluation. It can be seen from Figure 3.38 that as the number of nodes increases, the coefficient of variance increases, but it is always less than 1 (the value of the Poisson process). Thus, approximating the jittered VP cell traffic as a Poisson process yields conservative cell loss rates.

The above analysis indicates that the allowable traffic load (maximum link utilization) for a specific buffer size at a cross-connect node and a specific cell loss rate falls between the values for VPs without CDV, as determined by exact analysis [96,97] (shown in Figure 3.37 for VPs with the same T_0), and that for Poisson traffic (also in Figure 3.37) as determined by $M/D/1$ model analysis. In other words, the drop in the maximum link utilization as determined by the $M/D/1$ model is better for the case of multiplexing CBR VPs with CDV than without CDV. The drop approaches zero when each VP bandwidth ($1/T_0$) approaches zero or the number of nodes traversed approaches infinity (the jittered CBR-VP traffic approaches Poisson traffic).

The benefits of using the Poisson approximation in the analysis of jittered CBR VP multiplexing characteristics are summarized below.

1. The multiplexing characteristics become easy to analyze (the $M/D/1$ model can be used).
2. It allows the maximum link utilization to be determined irrespective of the attributes of each VP (peak cell rate, the number of nodes traversed, and the traffic conditions at each cross-connect node traversed) accommodated within the link. In other words, the approximation yields the minimum value of the link utilization for any combination of VP conditions.
3. The error caused by the approximation is insignificant for practical multiplexing node buffer sizes (usually more than 256 per outgoing port).
4. It yields a conservative estimate of the cell loss rate (converse statement of 2).
5. It enables efficient end-to-end cell-delay jitter evaluation as will be explained later in this section.

Cell Loss Priority

ATM cell headers contain a cell loss priority (CLP) bit for explicit cell loss priority indication, as mentioned in Section 3.2.2. This bit is used to indicate low-priority cells (CLP = 1), which can be discarded depending on network congestion. Using this bit or an implicit representation of VPI values, it is possible to provide different classes of cell loss for different VPs. This capability makes it possible to enhance network utilization because link utilization and the cell loss rate can be traded off against each other.

For the CBR VP applications discussed above, the increase in link utilization due to introducing VPs with higher acceptable cell loss rates was shown to be insignificant. Figure 3.39 shows the dependency of maximum link utilization on cell loss rate with buffer size as a parameter. All calculations were made with the $M/D/1$ model. It was found that when the buffer size is 256, the maximum allowable load is 0.99 and 0.96 for the specific cell loss rates of 10^{-2} and 10^{-9}, respectively. The difference is only 3%. If the traffic loads carried by high-quality VPs (10^{-9}) and low-quality VPs (10^{-2}) are equal, the expected gain in link utilization is only 1.5%. This result suggests that reasonable charge reductions stemming from the link utilization gain attained with the introduction of cell priority control are very small. On the other hand, the realization of multiclass QOS values requires selective cell handling at cross-connect nodes. This increases node hardware and control complexity which, in turn, increases node cost. If the link utilization gain stemming from multiclass QOS provisioning is sufficient to make up for the increased hardware/control cost, the introduction of multiclass QOS is justified. However, for CBR VP applications, the link utilization gain is usually too small to permit this. VP cross-connect systems providing only the most stringent cell loss rate are, therefore, very effective.

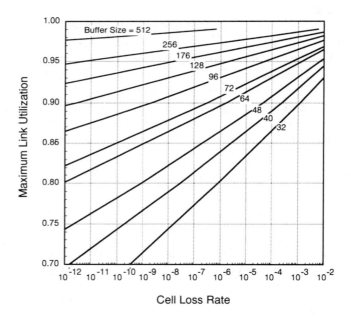

Figure 3.39 Maximum link utilization dependency on cell loss rate. (*Source:* [91]. Reprinted with permission.)

Cross-Connect Buffer Size Evaluation

The proposed approximation procedure was used to determine the buffer size needed for a subscriber line termination (SLT)/XC/ADM system. Figures 3.37 and 3.39 can be utilized for this purpose. The major observations from Figure 3.37 are as follows.

- Link utilization significantly increases with buffer size for buffer sizes less than 200.
- Link utilization is almost independent of buffer size if the buffer size is larger than 300.

From this, and from the cell-delay considerations described later in this section, the minimum buffer size (Q) of 256 was adopted for the systems developed for commercial use [93,102]. When Q is 256, and the specific cell loss rate is 10^{-10} (10^{-12}), the maximum link utilization efficiency is 0.956 (0.94).

End-to-End Cell-Delay Distribution

As discussed before (see Figure 3.38), the coefficient of variance of a CBR VP cell stream increases from 0 to a value less than 1 when the VP traverses tandem queuing nodes (cross-connect nodes). Exact analysis of the end-to-end cell-delay distribution on a VP must take into account the delay correlation of consecutive cells on the VP. Moreover, the exact output cell stream of each node must be strictly evaluated to derive the tandem queuing delay. However, it is impossible to achieve this in practice.

Here is shown another approximation method that allows the tandem queuing delay to be determined fairly easily. As described before, at each cross-connect node, cell traffic from other links decreases the consecutive cell-delay correlation, and if the cell traffic from the other links is about 50% that of the target input traffic, it has been shown that the cell-delay correlation can be neglected [100,101]. This was verified by computer simulations that examined various parameter values. With this assumption, the cell-delay distribution of a VP traversing a number of tandem nodes can be calculated by convoluting the cell-delay distribution at each node. As described before for CBR VPs, the upper bound of the cell-delay distribution at each node can be derived by the $M/D/1$ model. Thus the upper bound of the tandem queuing delay is given by the convolution of the cell-delay distribution obtained with the $M/D/1$ model.

It has been demonstrated that when traffic originating from other links is so small that the correlation still remains, the percentile tandem queuing delay of a VP cell is smaller than that obtained by a convolution that assumes there is no correlation in cell delay [100,101]. If all VPs are CBR VPs in the network concerned, it can be concluded that for any input traffic condition at any cross-connect/multiplexing nodes along the path, it is possible to calculate the upper bound of end-to-end cell delay on the path by convoluting the delay distribution at every node derived by $M/D/1$ analysis. Some evaluation examples are presented below.

Figure 3.40 shows network configuration examples for the evaluation of end-to-end cell-delay jitter (tandem queuing delay distribution). The evaluations assume the cell buffer size at each node to be 128 cells and the maximum utilization of each link to be 0.92 to guarantee a cell loss rate of better than 10^{-9} at each node. Model (a) in Figure 3.40 consists of a single transmission link speed (156 Mbps) while model (b) contains 156-Mbps and 2.5-Gbps transmission links. Figure 3.41 shows the 10^{-9} quantile cell delay (Pr[delay > 128 cell] = 10^{-9}) for the VP versus the number of nodes traversed as calculated by the proposed method for model (a). It is shown in Figure 3.41 that the delay incurred by the first node is the largest, and that the increase in cell delay decreases as the number of nodes traversed increases. This evaluation of tandem delay provides conservative values, and is much more effective than that obtained from a simple summation (instead of utilizing convolution) of the maximum delay of each node. For example, the simple summation scheme yields a maximum of $12 \times 20 = 2{,}560$ cells of tandem delay if the VP passes 20 nodes, while this method indicates only 360 cells of delay.

Figure 3.42 shows the 10^{-9} quantile cell delay for the VP in model (b). It is shown in Figure 3.42 that the cell delay that occurs on the 150-Mbps links at both ends is much larger than that on the 2.4-Gbps links. This method can easily be applied to networks consisting of links with a variety of transmission link speeds.

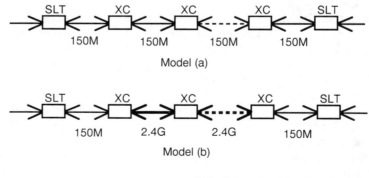

Figure 3.40 Network configuration examples.

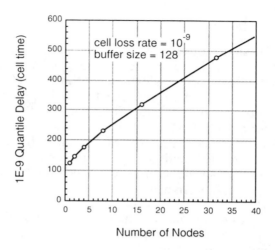

Figure 3.41 Tandem 1E-9 quantile delay for model (a). (*Source:* [91]. Reprinted with permission.)

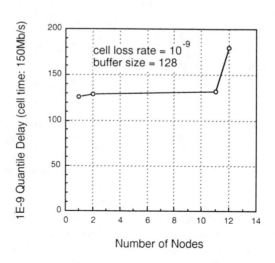

Figure 3.42 Tandem 1E-9 quantile delay for model (b). (*Source:* [91]. Reprinted with permission.)

VP Accommodation Design

VP connection admission control (CAC) is carried out by determining whether the VP connection can be accommodated while satisfying the required QOS on all links the VP will traverse. As explained, the CAC of a CBR VP can be effectively performed by applying the M/D/1 model approximation. This is done by estimating the maximum traffic load that satisfies both the cell-loss rate and delay requirements for the VP. That is, the traffic is determined by the minimum of ρ_L or ρ_D. Here, ρ_L is the maximum traffic load that satisfies the cell-loss rate requirement, and ρ_D is that which guarantees the cell-delay requirement. Details of this process are explained below.

The following notations are used:

- N: allowable maximum number of nodes traversed by the VP;
- β: specified end-to-end cell loss rate;
- D: specified end-to-end 10^{-m} quantile delay (cell time);
- Q: buffer size in each node.

Here, N is determined considering a transport network architecture consisting of backbone and regional transport networks. The cell loss due to a bit error in the cell header was assumed to be much smaller than that stemming from node buffer overflow.

First, we calculate ρ_L, which is determined from β. The cell-loss rate, α, assigned to each node is determined by the following equation:

$$(1 - \alpha)^N = 1 - \beta \tag{3.1}$$

From this equation, α is obtained as

$$\alpha \approx \beta/N \tag{3.2}$$

From Figure 3.39, the maximum traffic load ρ_L is determined to be that which guarantees the cell loss rate α of each node and hence guarantees the end-to-end cell-loss rate of β.

Next, ρ_D is determined from the end-to-end delay requirement. Figure 3.43 shows the relation of traffic load and 10^{-9} quantile cell delay where the number of traversed nodes is varied from 1 to 32. The evaluation was performed using the procedure described before. Here, m was assumed to be 9, and network model (a) was assumed for simplicity. From Figure 3.43, we can find ρ_D.

Finally, the maximum allowable traffic load $\rho_{\max}$, which guarantees both the cell-loss rate and the delay requirement is obtained as

$$\rho_{\max} = \min(\rho_L, \rho_D) \tag{3.3}$$

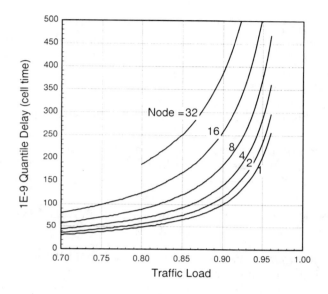

Figure 3.43 Tandem 1E-9 quantile delay. (*Source:* [91]. Reprinted with permission.)

Table 3.4 shows two example solutions obtained by this method. The specification of end-to-end delay for Example 1 is more severe than that of Example 2. In Example 1, ρ_{max} is restricted by ρ_D. This means that the buffer size Q at each node can be reduced until ρ_L becomes 0.902 (= ρ_D). In

Table 3.4
VP Accommodation Design Example

	Parameters	Example 1	Example 2
Design conditions	Maximum number of nodes, N	20	
	End-to-end cell loss rate, b	10^{-9}	
	Buffer size, Q	128	
	End-to-end delay, D	300 cell	400 cell
Cell loss rate at each node, α		5×10^{-11}	
	ρ_L	0.910	0.910
	ρ_D	0.902	0.925
	ρ_{max}	0.902	0.910

After: [91].

Example 1, Q can be reduced to 120 cells. In Example 2, on the other hand, ρ_{max} is restricted by ρ_L. If the buffer size Q can be increased, ρ_{max} can be increased up to 0.925 (= ρ_D). The important point to be noted here is that the specific cell-loss rate or cell delay occurs when all the conditions shown below are met simultaneously:

1. The number of nodes that the path traversed is the specified maximum value.
2. Utilization of all the links that the path traverses is the maximum value specified; in other words, each link fully accommodates VPs.
3. Utilization of all the above VPs is the maximum value specified; in other words, each VP fully accommodates VCs.
4. All the VCs accommodated within each VP above always launch cells at the maximum rates agreed at CAC.

It is impossible to evaluate the probability that all the above conditions will be met, but it is true that the probability is negligibly low in a practical sense. Even if it occurs, the VP accommodation design described here gives conservative QOSs. In most cases, therefore, much better performances are seen than the specific QOS values. Thus, the method explained here enables effective network QOS design, or VP accommodation design in an ATM network. The procedure can be used to control cell-delay variations by controlling link utilization.

This section described approximation methods for bandwidth allocation and QOS evaluation that are applicable to CBR VPs/VCs. It was shown that by adopting the Poisson arrival approximation, a complicated analytical method can be simplified and made effective for practical use; VP attributes such as peak cell rate and the number of nodes traversed, and the traffic conditions (conditions of other VPs) at each node are effectively eliminated from the analysis. Nevertheless, the approximation error has been shown to be small and conservative estimations of QOS are yielded. The methods can also evaluate end-to-end cell delay and the CDV, required buffer size for smoothing the CDV, and the CDV required to be considered for UPC/NPC, among other things.

3.4.4 VBR Path Accommodation Design

As explained in Section 3.4.3, the network-network VPs will mostly use CBR VPs, and so the major VPs in the network will CBR VPs. However, VBR VPs can be utilized in order to effectively utilize network resources or to capitalize on statistical multiplexing gain. The VBR VPs will mostly be utilized in the user-network VPs for leased line services [102]. VBR VP accommodation design is more difficult than that of CBR VP. A method that strictly guarantees the

QOSs in all conditions has not yet been developed, however, one practical and efficient method is introduced here [82,84,103–105].

The method utilizes the *worst UPC output pattern*, which depends on the UPC scheme adopted, as explained in Section 3.4.1. VP bandwidths can be conservatively allocated using the worst pattern. The bandwidth is calculated using the approximated cell-loss rate. This rate is determined from the expected cell arrival number based on the worst UPC output pattern using a time window equivalent to the buffer size. The path admission method is described below, and the flow is shown in Figure 3.44. Here, K is the buffer size, and i is the number of multiplexing VPs. The VP admission process is as follows:

1. Identify the worst cell pattern through the specific UPC circuit using the negotiated traffic descriptor values.
2. Calculate the probability distribution of the number of cells arriving over cell time K, $Pr[N_i(k) = x]$ for each VP_i ($i = 1, 2, \ldots$), based on the worst pattern. The value $Pr[N_i(k) = x]$ for each VP_i can be derived as follows:

For VP_i, derive the cell arrival function $f_i(t)$, which is defined below, for the worst cell arrival pattern:

$$f_i(t) = 1; \text{ if an information cell arrives} \quad (3.4)$$

or 0 if an empty cell arrives where t is normalized by the cell transfer time from the buffer.

Then, let N_t be the number of cells arriving during time interval $[t, t + K)$, and calculate $n_i[N_t = x]$, the frequency that N_t should equal x in the range $0 \leq t \leq T - 1$. Then derive the probability distribution, $Pr[N_i(k) = x]$ as

$$Pr[N_i(k) = x] = \frac{n_i[N_t = x]}{T} \quad (3.5)$$

3. Derive the probability distribution of the multiplexing state, $Pr[N(k) = x]$, by taking the convolution of all $Pr[N_i(k) = x]$ ($i = 1, 2, \ldots$)

$$Pr[N(k) = x] = Pr[N_1(k) = x] \otimes \cdots \otimes Pr[N_m(k) = x] \quad (3.6)$$

4. Evaluate approximated cell-loss probability from the equation

$$P[\text{loss}] = \frac{1}{E[N(K)]} \sum_{x > K} (x - K) \cdot Pr[N(k) = x] \quad (3.7)$$

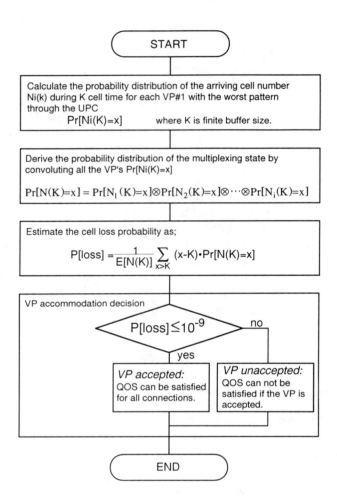

Figure 3.44 VBR path accommodation algorithm (*After:* [105]).

This provides the upper bound for cell-loss probability (conservative values for cell-loss probability). The proof is given in [106]. As is apparent from the above, this process can be easily executed, even though traffic descriptor values of each VP may differ.

5. Determine whether the VP can be accepted or not. If the calculated cell-loss rate from step 4 is larger than the specified QOS, the path is rejected.

The VP CAC procedure explained above is based on the worst cell pattern through the UPC and guarantees the conservative cell-loss rate at multiplexing.

If the worst traffic pattern cannot be strictly determined, however, as in the case of LB algorithm, the conservativeness cannot be assured. Furthermore, the above process only takes account of one multiplexing node; that is, the effect of cell multiplexing at the following nodes along the VP is not considered. Unlike the CBR VP accommodation design explained before, the squared coefficient of variation of cell interval for a VBR VP can be more than 1. Even if it is less than 1, it can exceed 1 (the value for random traffic) after traversing a number of cross-connect nodes. This occurs when the VP is multiplexed with other bursty VPs along the node. Therefore, in VBR VP accommodation, the upper value of VP burstiness cannot be bounded, so the Poisson approximation cannot be applied. In VBR VP accommodation, therefore, some congestion control mechanisms must be used with the above process in order to strictly guarantee the QOSs.

3.4.5 Evaluations of Network Utilization Enhancement

One of the important techniques to reduce the transport cost of a connection is to maximize the utilization of network resources. This is attained by capitalizing on the statistical multiplexing effect as explained in Section 3.4.2. Here, examples are shown of the statistical multiplexing gain obtained with burst-level and cell-level resource sharing principles. The former is used by reserved mode connections, where the VP bandwidth can be altered according to user demand. When a user demand for bandwidth increase is accepted by the network, the QOS of the VP is guaranteed by the network. Decreases in the bandwidth are allowed at any time. The improvement of the link utilization or resultant cost reduction is offset by the introduction of nonzero blocking probability for bandwidth increase requests and by the increased processing load incurred by the on-demand resource reservation.

Cell-level statistical multiplexing gain is obtained by utilizing VBR VPs. Some calculation results of these two levels of statistical multiplexing are demonstrated here.

Reserved-Mode Connection

Figure 3.45(a) explains the basic concept for variable-capacity VP connections. The minimum VP bandwidth for the connection is W_1 (can be 0), and the user can increase the bandwidth by W_2 on demand when there is sufficient unused link capacity along the VP end to end. Here, CBR VPs are considered for simplicity (the peak rates of W_1 and W_2 are constant). Charging is based on W_1 and W2 and on the duration of each bandwidth, T_1 and T_2, as shown in Figure 3.45(a).

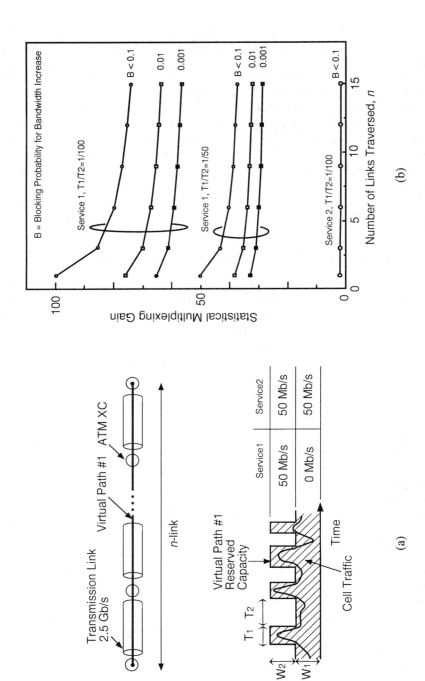

Figure 3.45 Reserved-mode VP connection: (a) basic concept for variable capacity VP services and (b) VP capacity control effects (After: [102]).

Figure 3.45(b) shows analytical results on the statistical gain for a VP traversing n links. The statistical gain is defined here as the ratio of the maximum allowable number of VPs with the reserved-mode scheme to that with the permanent-mode scheme (peak rate is constant at $W_1 + W_2$). It was found that the worst blocking probability for a VP that traverses n links occurs when all other VPs traverse only one link [62]. Details of the analytical procedures are given in [62,63]. The probabilities of failure to increase the bandwidth (B - blocking probability; see Figure 3.45(b)) were set at 0.1 and 0.01. Link speed was set at 2.5 Gbps, and the values of W_1 and W_2 are shown in the figure. The duration of T_i was assumed to have a negative exponential distribution with the average value of T_1 or T_2, and the ratio of T_1/T_2 was set at 1/50 and 1/100. The point to be emphasized is that statistical gain is only slightly dependent on n, and that a gain of more than 50 can be attained when T_1/T_2 is 1/100.

Statistical Cell Multiplexing Gain

Figure 3.46 shows an example of the statistical multiplexing gain obtained with the algorithm described in Section 3.4.4, where T_0 is the minimum cell interarrival time to a VP and T is the period defining the average cell rate. The buffer size K is set at 128. All the VPs multiplexed are assumed to have the same bandwidth parameter values. The gain is defined as the reciprocal of the required bandwidth normalized to that required with peak cell rate bandwidth allocation. Figure 3.46 shows that gain of more than 10 is obtained depending on VP bandwidth parameter values.

3.5 NETWORK ELEMENT

An organic network architecture of ATM networks is depicted in Figure 3.47. The customer premises network (CPN) will utilize an ATM multimedia multiplexer, which provides service-dependent terminal interfaces for ATM LAN, ATM switched services, ATM leased line services, and ATM video distribution services. It terminates VPs and will also terminate VCs and multiplex them into different VPs. Subscriber line termination (SLT) will be implemented in the service nodes. VP UPC function is implemented in SLT and, after the UPC, VPs from different customers are multiplexed and fed to the ATM cross-connect system. The ATM cross-connect systems route the cells according to their VPIs. The ATM cross-connect systems can be interlinked directly by the SDH transmission systems of STM-N (cells are mapped into VC-4-Nc (N = 4, 16)) or by STM-1 according to the traffic demand. ATM to STM conversion will be performed if necessary to transmit ATM signals utilizing STM LT-MUX in conjunction with STM signals. In the subscriber access network, ATM ADM

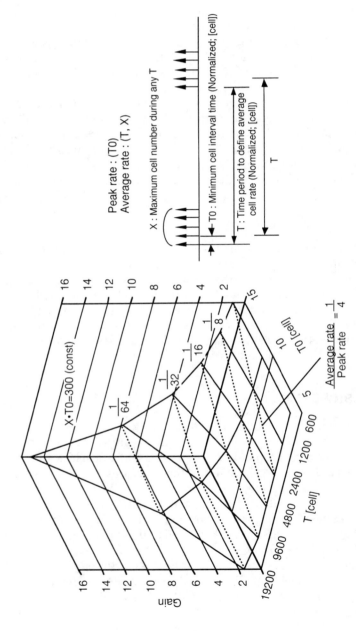

Figure 3.46 Statistical cell multiplexing effect (*After:* [102]).

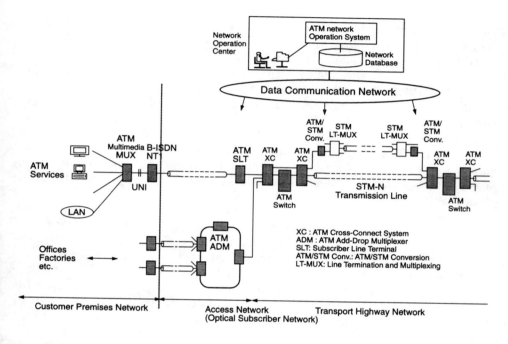

Figure 3.47 Organic architecture of ATM-based network.

systems forming a physical ring structure may be applied [94,95] depending on customer distribution and density. Some of these key hardware systems necessary to create ATM transport networks are explained below.

3.5.1 Cross-Connect System

To implement the VPs that can realize the benefits explained in Section 3.3, the key network element (NE) is the ATM cross-connect system. For cross-connection, paths/channels in different lines have to be interchanged. In the process, the multiplexing data width has an impact on system implementation issues. Existing PDH cross-connect systems are based on the bit multiplexing technique. SDH employs byte multiplexing. ATM utilizes cell (53 bytes) multiplexing. If we are creating a large-bandwidth network, the capacity of the cross-connect systems should be very large. To attain this, data speed within the ATM cross-connect system should be enhanced by utilizing parallel processing up to the cell-length. This allows the line capacity supported by the system to be increased. The SDH cross-connect system can employ only byte-parallel techniques. The PDH cross-connect system is basically serial, where the higher speed line is demultiplexed into lower speed lines, and the lower speed line is interchanged by manual wiring or by using a space division switch. Thus, it is difficult to realize a large-capacity PDH cross-connect network economically.

For the ATM switch fabric, various switch architectures have been proposed [107,108]. The architectures can be characterized into switch topology (matrix, bus, ring), buffer configuration (output, input, shared, crosspoint), and cell-processing principles (FIFO, FIRO, RIRO, scheduling). The most representative topology is the matrix type, and these switches are classified in Table 3.5 in terms of buffer configuration. Here, input and output port numbers are N. In the output buffer type switch, input cells are multiplexed onto a high-speed bus in order to prevent the cell collisions that would otherwise occur when cells simultaneously arriving at different input ports are destined to the same output port. Cells are written into the buffer at N times the link speed. The necessary buffer size can be evaluated considering link utilization, cell arrival pattern of each VP/VC, switch size, and the specified cell-loss probability of the fabric (see Section 3.4). In the shared-buffer-type switch, buffers are shared among N input ports, and therefore, the necessary total buffer quantity can be reduced compared to the output buffers. The larger the switch size, the larger the reduction. On the other hand, to share buffers, cell reading must be high-speed and buffer management is necessary. In the crosspoint-buffer-type switch, buffers are allocated at each crosspoint. Reading and writing of cells from and into buffers can be done at link speed. It is easy, therefore, to accommodate high-speed links; however, N^2 buffers are necessary, which limits the development of large-scale switches. Additional switch functions such as broadcasting and cell priority control can be implemented easily with the output and crosspoint-buffer-type switches. These buffer allocation approaches can be combined to optimize switch performance against specific requirements.

The requirements of the ATM cross-connect system are as follows:

- Different interface speeds must be efficiently accommodated.
- Modular growth capability must be ensured to allow economical introduction of systems to large or small offices.
- They must offer large throughput and minimum delay.

An example of such cross-connect system architecture [54,111] is shown in Figure 3.48. The system utilizes a 2.5-Gbps operating speed switch fabric. This is because the cross-connect system is designed to accommodate high-speed interfaces effectively and to minimize cell buffer delay. The inner hardware operation speed is reduced by utilizing parallel processing techniques in which the number of parallel lines can be up to the cell length. The requirements for the ATM cross-connect switch fabric are as follows:

Table 3.5
Different ATM Switch Types

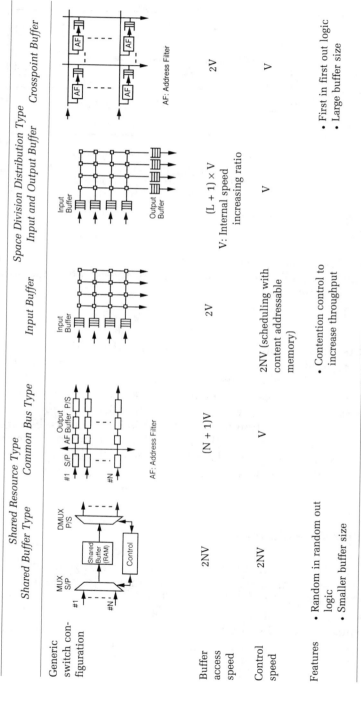

	Shared Resource Type		Space Division Distribution Type		
	Shared Buffer Type	Common Bus Type	Input Buffer	Input and Output Buffer	Crosspoint Buffer
Generic switch configuration	(diagram)	(diagram)	(diagram)	(diagram)	(diagram)
Buffer access speed	2NV	$(N+1)V$	2V	$(L+1) \times V$ V: Internal speed increasing ratio	2V
Control speed	2NV	V	2NV (scheduling with content addressable memory)	V	V
Features	• Random in random out logic • Smaller buffer size		• Contention control to increase throughput		• First in first out logic • Large buffer size

After: [108].

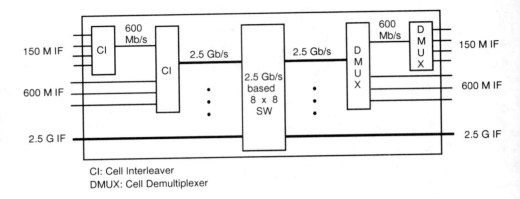

CI: Cell Interleaver
DMUX: Cell Demultiplexer

Figure 3.48 Generic cross-connect system architecture. (*After:* [111]).

- Nonblocking characteristics in a strict sense;
- No restrictions on the usage condition of the middle link in multistage switching networks;
- Cell sequence integrity.

The first requirement is necessary to enable effective VP restoration, which does not require any rearrangement if output port bandwidth is available for restoration paths. The second requirement means that the switch connectivity should be determined only by input and output port connection relationships, and should not be restricted by middle link conditions. This allows the blocking stemming from the utilization conditions of the inner link to be avoided even when VP capacity is dynamically altered in the network, as explained in Section 3.3. Examples of switch fabrics that satisfy the above requirements are one-stage-switching networks or self-routing multistage networks.

3.5.2 Other Network Elements

At ATM introduction, an ATM cross-connect system interfacing with SDH-based transmission networks will be introduced to provide VPs between existing STM-based switching systems. STM-ATM signal conversion [109,110] (see Figure 3.12) is needed to connect SDH paths and VPs. Other important network elements include NTs, SLTs, and ADMs. The functional configurations of these elements are depicted in Figures 3.49 and 3.50 [112].

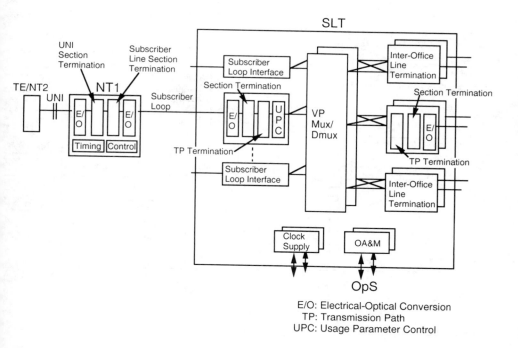

Figure 3.49 ATM NT and SLT functional configuration (*After:* [112]).

3.6 CONCLUSION

The inherent benefit of ATM technology, if we refer to it in one word, is the flexibility that it provides. This is made possible by virtue of the unique information transfer unit of the cell and of the store-and-forward transfer principle. This, however, entails rather sophisticated cell traffic management technologies to maximally exploit its potential. Therefore, the basics of ATM have been established, while detailed standards continue to evolve, especially for traffic management and OA&M-related issues.

ATM technology was intended to be the basis on which all services were supported, so a relatively short cell length was adopted to reduce cell assemble delay for the telephone service. ATM technologies and systems are now being introduced to overlay existing PDH and SDH networks, mainly for high-speed data services. Eventually, all telephone and narrowband data traffic will be transferred using ATM, but due to the large investment needed, it will take many years—maybe more than a decade—before this occurs. The transition will be spurred by the substantial increase in high-speed data and video traffic

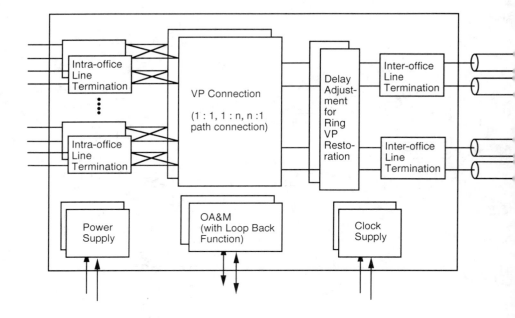

Figure 3.50 ATM-ADM functional configuration (*After:* [112]).

expected, but that is possible only after very cost effective end-to-end large bandwidth transport capabilities become available. The key technology to enable this—the photonic network—will be discussed in the next chapter.

References

[1] Inoue, Y., I. Tokizawa, and N. Terada, "Basic consideration to define broadband network interfaces," *Proc. GLOBECOM 87*, Tokyo, Japan, Nov. 15–18, 1987, pp. 13.2.1–5.

[2] Beckner, M. Wm., T. T. Lee, and S. E. Minzer, "A protocol and prototype for broadband subscriber access to ISDN's," *Proc. ISS*, Phoenix, AZ, March 15–20, 1987.

[3] Kultzer, J. K., and W. A. Montgomery, "Statistical switching architecture for future services," *Proc. ISS 84*, Florence, May 7–11, 1984, Session 43 A, Paper 1, pp. 1–6.

[4] Turner, J. S., "New directions in communication," *Proc. Int. Zurich Seminar on Digital Commun.*, Zurich, March 11–13, 1986, pp. A3.1–A3.7.

[5] Hagimoto, K., S. Nishi, K. Nakagawa, "An optical bit-rate flexible transmission system with 5-Tb/s.Km capacity employing multiple in-line erbium-doped fiber amplifiers," *IEEE J. Lightwave Technology*, Vol. 8, No. 9, 1990, p. 1387.

[6] Ishio, H., "Progress of fiber optic technologies and their impact on future telecommunication networks," *Proc. FORUM '91* (Technical Symposium of TELECOM 91), Geneva, Switzerland, Oct. 10–15, 1991, Session 2.2, pp. 117–120.

[7] Nakagawa, K., K. Aida, K. Aoyama and K. Hohkawa, "Optical amplification in trunk transmission networks," *IEEE LTS*, Vol. 3, No. 1, Feb. 1992, pp. 19–26.

[8] Tokizawa, I., T. Kanada, and K. Sato, "A new transport network architecture based on Asynchronous Transfer Mode techniques," *Proc. ISSLS 88*, Sept. 11–16, 1988, Boston, MA, pp. 11.2.1.–11.2.5.
[9] Requirements for interfacing digital terminal equipment to services employing the extended superframe format, PUB54010, AT&T, Oct. 1984.
[10] Maeda, Y., M. Tokunaga, and I. Tokizawa, "An advanced multimedia TDM system for closed network," *Proc. ICC 87*, 1987, pp. 30A.3.1–5.
[11] Okano, Y., T. Kawata, and T. Miki, "Designing digital paths in transmission networks," *Proc. GLOBECOM 86*, Houston, TX, 1986, pp. 25.2.1–25.2.5.
[12] Okano, Y., S. Ohta, and T. Kawata, "Assessment of cross-connect systems in transmission networks," *Proc. GLOBECOM 87*, Tokyo, Japan, Nov. 15–18, 1987.
[13] Roberts, L., "The Evolution of Packet Switching." *Proc. IEEE*, Vol. 66, Nov. 1978, pp. 1307–1313.
[14] Kleinrock, L., "Principles and Lessons in Packet Communications," *Proc. of the IEEE*, Vol. 66, No. 11, Nov. 1978, pp. 1320–1329.
[15] Folts, H. C., "Interface Standards for Public Data Networks," *National Communication System Technical Information Bulletin*, 79-2, March 1979.
[16] Rosner, R. D., *Packet Switching*, Belmont, CA: Lifetime Learning Publications, 1982.
[17] Turner, J. S., and L. F. Wyatt, "A Packet Network Architecture for Integrated Services," *Proc. Globecom '83*, 2.1.1–2.1.6, Nov.–Dec. 1983.
[18] Turner, J. S., "Design of an Integrated Services Packet Network," *IEEE Journal on Selected Areas in Communication*, Vol. SAC-4, No. 8, Nov. 1986, pp. 1373–1380.
[19] Turner. J. S., "Design of a Broadcast Packet Network," *Proc. INFOCOM '86*, 1986, pp. 667–675.
[20] Turner, J. S., "New Directions in Communications (or Which Way to the Information Age?)," *IEEE Communications Magazine*, Vol. 24, No. 10, Oct. 1986, pp. 8–15.
[21] Haselton, E. F., "A PCM Frame Switching Concept Leading to Burst Switching Network Architecture," *IEEE Communications Magazine*, Sept. 1983, pp. 13–19.
[22] Amstutz, S., "Burst Switching-An Introduction," *IEEE Communications Magazine*, Nov. 1983, pp. 36–42.
[23] O'Reilly, P., "Burst and Fast-Packet Switching: Performance Comparisons," *Proc. INFOCOM '86*, 1986, pp. 653–666.
[24] Devault, M., J. P. Quinquis, and Y. Rouaud, "-Asynchronous time division switching: A new concept for ISDN nodes," *Proc. ISS '81*, Session 42B, Paper 5, Montreal, Sept. 21–25, 1981.
[25] Devault, M., D. Chomel, H. Le Bris, and Y. Rouaud, "From data to moving pictures: A multi-bit rate asynchronous time-division equipment at the subscriber's premises," *Proc. ISSLS '84*, Nice, Sept. 1984, pp. 283–287.
[26] Thomas, A., J. P. Courdreuse, and M. Servel, "Asynchronous time-division techniques: An experimental packet network integrating video communication," *Proc. ISS '84*, Session 32C, Paper 2, Florence, May 7–11, 1984.
[27] Gonet, P., P. Adam, and J. P. Coudreuse, "Asynchronous time-division switching: The way to flexible broadband communication networks," *Proc. Int. Zurich Seminar on Digital Commun.*, Zurich, March 11–13, 1986, pp. D5.1–D5.7.
[28] CCITT SG XVIII, Delayed Contribution COM XVIII D.414/XVIII, Definition of a new bearer service using Asynchronous Time Division techniques, Kyoto, Dec. 2–13, 1985, (Source: France).
[29] CCITT SG XVIII, Delayed Contribution COM XVIII FR26, The Asynchronous Time Division transfer mode: Some possible technical features. Answer to the questionnaire of the Kyoto Meeting (Annex 7 to part V, WP XVIII/1 - R12), Geneva, July 1986.
[30] CCITT SG XVIII, ISDN Expert Meeting, Brasilia, Feb. 2–13, 1987, "Report of the Brasilia

meeting of broadband task group -Part B," TD130, (Source: Chairman Broadband Task Group of SG XVIII).
[31] CCITT SG XVIII, Hamburg Meeting, July 1–10, 1987, "Part C.9 of the report of working party XVIII 15: Proposed draft recommendation I.113 -Vocabulary of terms for broadband ISDNs," TD No. 35 (Source: Special rapporteur for question 19/XVIII).
[32] Sato, K., S. Ohta, and I. Tokizawa, "Broadband ATM network architecture based on Virtual Path, " *IEEE Trans. Commun.*, Vol. 38, No. 8, Aug. 1990, pp. 1212–1222.
[33] ITU-T Recommendation I.311, B-ISDN General Network Aspects, March 1993.
[34] ITU-T Recommendation I.361, B-ISDN ATM Layer Specification, March 1993.
[35] ITU-T Recommendation I.610, B-ISDN Operation and Maintenance Principles and Functions, March 1993.
[36] ITU-T Recommendation I.411, User-Network Interfaces -Reference Configurations, March 1993.
[37] ITU-T Recommendation I.413, B-ISDN User-Network Interface, March 1993.
[38] Sato, K., and H. Hoshi, "Digital Integrated Transport Network-II," *NTT R&D Document*, No. 697, May 1987.
[39] Kanada, T., K. Sato, and T. Tsuboi, " An ATM based transport network architecture," *Proc. IEEE COMSOC Int. Workshop on Future Prospects of Burst/Packetized Multimedia Commun.*, Nov. 22–24, 1987, Osaka, Japan.
[40] Sato, K., T. Kanada, and I. Tokizawa, "High-speed burst transport system architecture," Paper of Technical Group, IN87-84, IEICE Japan, Dec. 24, 1987.
[41] CCITT SG XVIII, Seoul Meeting, Jan. 25–Feb. 5, 1988, "Some key features on UNI and NNI considering ATM (A framework for discussion)," Delayed Contribution No. D.1566/XVIII, (Source: NTT).
[42] CCITT SG XVIII, Seoul Meeting, Jan. 25–Feb. 5, 1988, "ATM cell header functional requirements and length," Temporary Document WD8(BBTG)/XVIII, (Source: Australia).
[43] Addie, R. G., and R. E. Warfield, "Bandwidth switching and new network architecture," *Proc. 12th Int. Teletraffic Congress*, Turin, June 1988, 2.3ii A.1.1–A.1.7.
[44] Ohta, S., K. Sato, and I. Tokizawa, "A dynamically controllable ATM transport network based on the Virtual Path concept," *Proc. GLOBECOM 88*, Nov. 28–Dec. 1, 1988, Lauderdale, FL, pp. 39.2.1.–39.2.5.
[45] Foster, G., and J. L. Adams, "The ATM zone concept," *Proc. GLOBECOM 88*, Nov. 28–Dec. 1, 1988, Lauderdale, FL, pp. 21.4.1.–21.4.3.
[46] Sato, K., S. Ohta, and I. Tokizawa, "Dynamic reconfiguration of ATM-based network via Virtual Path strategy," *Presented at the 2nd IEEE COMSOC International Multimedia Communications Workshop*, Ottawa, Canada, April 20–23, 1989.
[47] Coudreuse, J-P., and L. Etesse, "ATM: Status of definition and discussion of some open issues," *Presented at the 2nd IEEE COMSOC International Multimedia Communications Workshop*, Ottawa, Canada, April 20–23, 1989.
[48] Sato, K., S. Ohta, and I. Tokizawa, "Broadband ATM network architecture based on Virtual Path, " *IEEE Trans. Commun.*, Vol. 38, No. 8, Aug. 1990, pp. 1212–1222.
[49] Sato, K., S. Ohta, and I. Tokizawa, "Broadband transport network architecture based on Virtual Path concept," *Trans. on IEICE Japan*, Vol. J72-B-I, No. 11, Nov. 1989, pp. 904–916.
[50] Sato, K., and I. Tokizawa, "Flexible asynchronous transfer mode networks utilizing Virtual Paths," *Proc. ICC '90*, Atlanta, GA, April 16–19, 1990, pp. 318.4.1.–318.4.8.
[51] Tirtaatmadja, E., and R. A. Palmer, "The application of virtual paths to the interconnection of IEEE 802.6 metropolitan area networks," *Proc. ISS '90*, Stockholm, Vol. II, May 27–June 1, 1990, pp. 133–137.
[52] Sato, K., H. Hadama, and I. Tokizawa: "Network reliability enhancement with Virtual Path strategy," *Proc. GLOBECOM '90*, San Diego, Dec. 2–5, 1990, 403.5.

[53] Tokizawa, I., and K. Sato, "Network reliability enhancement with Virtual Path strategy," *Proc. GLOBECOM '90*, San Diego, Dec. 2–5, 1990, pp. 705B.4.1–5.
[54] Sato, K., H. Ueda, and N. Yoshikai, "The role of Virtual Path cross-connection," *IEEE LTS* (The Magazine of Lightwave Telecommunication Systems), Vol. 2, No. 3, Aug. 1991, pp. 44–54.
[55] Coudreuse, J-P., "Network evolution towards BISDN," *IEEE LTS*, Vol. 2, No. 3, Aug. 1991, pp. 66–70.
[56] CCITT I-series Recommendations (B-ISDN), Nov. 1990.
[57] Shirakawa, H., K. Maki, and H. Miura, "Japan's network evolution relies on SDH-based systems," *IEEE LTS*, Nov. 1991, pp. 14–18.
[58] Kasai, H., T. Murase, and H. Ueda, "Synchronous digital transmission systems based on CCITT SDH standard," *IEEE Communications Magazine*, Vol. 28, No. 8, Aug. 1990, pp. 50–59.
[59] Sato, K., R. Kawamura, and I. Tokizawa, "Introduction strategy of B-ISDN based on virtual paths," *Proc. FORUM '91* (Technical Symposium of TELECOM 91), Geneva, Switzerland, Oct. 10–15, 1991, Session 2.5, pp. 225–229.
[60] CCITT SG XVIII, Geneva Meeting, June 1989, Leased line service implementation techniques and the necessity of freedom to implement user access capability to virtual paths at the UNI, Delayed Contribution No. D.313/XVIII, (Source: NTT).
[61] Ohta, S., and K. Sato, "Dynamic bandwidth control of the Virtual Path in an Asynchronous Transfer Mode network," *IEEE Trans. on Commun.*, Vol. 40, No. 7, July 1992, pp. 1239–1247.
[62] Hadama, H., K. Sato, and I. Tokizawa, "Dynamic bandwidth control of Virtual Paths in ATM networks," Presented at *3rd IEEE Int. Workshop on Multimedia Commun., MULTIMEDIA '90*, Bordeaux, France, Nov. 14–17, 1990.
[63] Hadama, H., K. Sato, and I. Tokizawa, "Analysis of Virtual Path bandwidth control effects in ATM networks," *IEICE Trans. on Commun.*, Vol. E78-B-I, No. 6, June 1995, pp. 907–915.
[64] Grover, W. D., "The self-healing network," *Proc. GLOBECOM '87*, Tokyo, Japan, Nov. 15–18, 1987, pp. 28.2.1–28.2.6.
[65] Hasegawa, S., A. Kanemasa, H. Sakaguchi, and R. Maruta, "Dynamic reconfiguration of digital cross-connect systems with network control and management," *Proc. GLOBECOM '87*, Tokyo, Japan, Nov. 15–18, 1987, pp. 28.3.1–28.3.5.
[66] Yang, C. H., and S. Hasegawa, "FITNESS-Failure immunization technology for network service survivability," *Proc. GLOBECOM '88*, Nov. 28–Dec. 1, Lauderdale, FL, pp. 47.3.1.–47.3.6.
[67] Amirazizi, H. R., "Controlling Synchronous Networks with Digital Cross-Connect Systems," *Proc. GLOBECOM '88*, Nov. 28–Dec. 1, Lauderdale, FL, pp. 1560–1560.
[68] Grover, W. D., B. D. Venables, J. H. Sandham, and A. F. Milne, "Performance Studies of a Selfhealing Network Protocol in Telecom Canada Long Haul Networks," *Proc. GLOBECOM '90*, San Diego, CA, Dec. 2–5, 1990, pp. 452–458.
[69] Sakauchi, H., Y. Nishimura, and S. Hasegawa, "A Self-Healing Network with an Economical Spare-Channel Assignment," *Proc. GLOBECOM '90*, San Diego, CA, Dec. 2–5, 1990, pp. 438–443.
[70] Komine, H., T. Chujo, T. Ogura, K. Miyazaki, and T. Soejima, "A Distributed Restoration Algorithm for Multiple-Link and Node Failures of Transport Networks," *Proc. GLOBECOM '90*, San Diego, CA, Dec. 2–5, 1990, pp. 459–463.
[71] Kawamura, R., K. Sato, and I. Tokizawa, "Self-healing network techniques utilizing Virtual Paths," *Proc. 5th Int. Network Planning Symposium, Networks '92*, Kobe, Japan, May 17–22, 1992, pp. 129–134.
[72] Kawamura, R., H. Hadama, and K. Sato, "Self-Healing Techniques Utilizing Virtual Path

Concept for ATM networks," *Electronics and Communications in Japan*, Part 1, Vol. 75, No. 4, 1992, pp. 86–96.

[73] Kawamura, R., K. Sato, and I. Tokizawa, "Self-healing ATM networks based on Virtual Path concept," *IEEE J-SAC special issue on Integrity of Public Telecommunication Networks*, Vol. 12, No. 1, Jan. 1994, pp. 120–127.

[74] Tokura, N., K. Oguchi, and T. Kanada, "A broadband subscriber network using optical star coupler," *Proc. GLOBECOM '87*, Nov. 1987, 37.1.

[75] Stern, J. R., et al., "TPON -a passive optical network for telephony," *Proc. 14th ECOC '88*, Brighton, 1988, pp. 203.

[76] McCarter, J., et al., "Development of passive optical networks for subscriber loop application in Australia," *Proc. IWCS*, Atlanta GA, 1989.

[77] ITU-T Recommendation I.371, Traffic Control and Congestion Control in B-ISDN, March 1993.

[78] Brochin, F. M., "A Cell Admission policy for LAN to LAN Interconnection," *IEICE Technical Report*, SAT91-11, 1991.

[79] Guillemin, F., P. Boyer, and L. Romoeuf, "The Spacer-Controller: Architecture and First Assessments," *Workshop on Broadband Communications*, Estoril, Portugal, Jan. 1992, pp. 294–305.

[80] Boyer, P. E., F. M. Guillemin, M. J. Servel, and J. P. Coudreuse, "Spacing Cells Protects and Enhances Utilization of ATM Network Links," *IEEE Network*, Vol. 6, No. 5, Sept. 1992, pp. 38–49.

[81] Yamanaka, N., Y. Sato, and K. Sato, "Traffic shaping for VBR traffic in ATM networks," *IEICE Trans. Commun.*, Vol. E75-B, No. 10, Oct. 1992, pp. 1105–1108.

[82] Yamanaka, N., Y. Sato, and K. Sato, "Usage parameter control and bandwidth allocation methods for ATM-based B-ISDN," *Presented at 4th IEEE COMSOC Int. Workshop, Multimedia '92*, Monterey, CA, April 1–4 1992, pp. 210–216.

[83] Yamanaka, N., Y. Sato, and K. Sato, "Precise UPC scheme suitable for ATM networks characterized by widely ranging traffic parameter values," *IEICE Trans. Commun.*, Vol. E75-B, No. 12, Dec. 1992, pp. 1367–1372.

[84] Yamanaka, N., Y. Sato, and K. Sato, "Usage parameter control and bandwidth allocation methods considering cell delay variation in ATM networks," *IEICE Trans. Commun.*, Vol. E76-B, No. 3, March 1993, pp. 270–279.

[85] Costipaiboon, R., and V. Phung, "Usage Parameter Control and Bandwidth Allocation Method for B-ISDN/ATM Variable Bit Rat Services," *Proc. Multimedia '90*, Session 4, Bordeaux, France, Nov. 1990.

[86] Yamanaka, N., Y. Sato, and K. Sato, "Performance limitation of leaky bucket algorithm for usage parameter control and bandwidth allocation methods," *IEICE Trans. Commun.*, Vol. E75-B, No. 2, Feb. 1992, pp. 82–86.

[87] Tranchier, D. P., P. E. Boyer, Y. M. Rouaud, and J-Y. Mazeas, "First Bandwidth Allocation in ATM Networks," *Proc. ISS '92*, A5.2, Japan, Oct. 1992.

[88] Watanabe, Y., and H. Hadama, "Adaptive Virtual Path capacity control in multimedia ATM networks, " *Trans. IEICE*, Vol. J76-B-I, No. 7, July 1993.

[89] Kawamura, R., H. Hadama, K. Sato, and I. Tokizawa, "Fast VP-bandwidth management with distributed control in ATM networks," *IEICE Trans. Commun.* Vol. E77-13, No. 1, Jan. 1994, pp. 5–14.

[90] Saito, H., *Teletraffic Technologies in ATM Networks*, Artech House: Norwood, MA, 1994.

[91] Sato Y., N. Yamanaka, and K. Sato, "ATM network resource management techniques for CBR virtual paths/channels," *IEICE Trans. Commun.* Vol. E79-B, No. 5, May 1996, pp. 684–692.

[92] Sato, K., R. Kawamura, and I. Tokizawa, "Introduction strategy of B-ISDN based on virtual

paths," *Proc. FORUM '91* (Technical Symposium of TELECOM 91), Geneva, Switzerland, Oct. 10–15, 1991, Session 2.5, pp. 225–229.

[93] Aoyama, T., I. Tokizawa, and K. Sato, "An ATM VP-based broadband network: its techniques and application to multimedia services," *Proc. ISS '92*, Yokohama, Japan, Oct. 25–30, 1992, A1.3.

[94] Tokura, N., K. Kikuchi, and K. Oguchi, "Subscriber access network architecture based on ATM techniques," *Proc. GLOBECOM '89*, Dallas, TX, Nov. 27–30.

[95] Maeda, Y., K. Kikuchi, and N. Tokura, "ATM access network architecture," *Proc. ICC 91*, Denver, CO, June 1991, 23.1.

[96] Eckberg, A.E., "The single server queue with periodic arrival process and deterministic service time," *IEEE Trans. Commun.*, Vol. 27, March 1979, pp. 556–562.

[97] Bhargava, A., P. Humblet, and M. G. Hluchy, "Queueing analysis of continuous bit-stream transport in packet network," *IEEE GLOBECOM '89*, 1989.

[98] Roberts, J. W., and J. T. Virtamo, "The superposition of periodic cell arrival streams in an ATM multiplexer," *IEEE Trans. Commun.*, Vol. 39, No. 2, pp. 298–303, Feb. 1991.

[99] Nakagawa, K., "Queueing analysis of CBR inputs with multiple periods," *Paper of Technical Group*, CS91-3, IEICE Japan, May 1991.

[100] Sato, K., H. Nakada, and Y. Sato, "Variable rate speech coding and network delay analysis for universal transport network," *Proc. INFOCOM '88*, New Orleans, LA, March 27–31, 1988, pp. 8A.2.1–8A.2.10.

[101] Sato, Y., and K. Sato, "Delay analysis of multi-node queueing system," *Trans. on IEICE Japan*, Vol. J71-B, No. 6, June 1988, pp. 669–677.

[102] Aoyama T., I. Tokizawa, and K. Sato, "Introduction strategy and technologies for ATM VP-based broadband network," *IEEE J-SAC*, Vol. 10, No. 9, Dec. 1992, pp. 1434–1447.

[103] Sato, Y., N. Yamanaka, K. Sato, and N. Tokura, "Experimental ATM transport system and virtual path management techniques," *Proc. GLOBECOM 91*, Phoenix, AZ, Dec. 2–5, 1991, Session 59, pp. 59.6.1–59.6.7.

[104] Sato, Y., and K. Sato, "Evaluation of statistical cell multiplexing effects and path capacity design in ATM networks," *IEICE Trans. Commun.*, Vol. E75-B, No. 7, July 1992, pp. 642–648.

[105] Yamanaka, N., Y. Sato, and K. Sato, "Precise UPC scheme and bandwidth allocation methods for ATM-based B-ISDN characterized by wide-ranging traffic parameter values," *Int. Journal of Digital and Analog Communication Systems*, Vol. 5, No. 4, Oct.–Nov. 1992, pp. 203–215.

[106] Saito, H., "New dimensioning concept for ATM networks," *Proc. International Teletraffic Congress Specialist Seminar*, Morristown, NJ, 1990.

[107] G-Hara, J., and A. Jajszczyk, "ATM shared-memory switching architectures," *IEEE Network*, Vol. 8, No. 4, July/Aug. 1994, pp. 18–26.

[108] Koinuma, T., and T. Takahashi, "ATM node system technologies," *NTT REVIEW*, Vol. 6, No. 1, Jan. 1994, pp. 43–48.

[109] Uematsu, H., and H. Ueda, "STM signal transfer techniques in ATM networks," *Proc. ICC '92*, Chicago, IL, June 14–18, 1992, pp. 311.6.1–311.6.5.

[110] Uematsu, H., and H. Ueda, "Implementation and experimental results of CLAD using SRTS method on ATM networks," *Proc. GLOBECOM '94*, San Francisco, CA, Nov. 29–Dec. 2, 1994, pp. 1815–1821.

[111] Ueda, H., H. Obara, H. Uematsu, and H. Ohta, "ATM cross-connect system technology," *NTT R&D*, Vol. 42, No. 3, 1993, pp. 357–366.

[112] Tokura, N., T. Tsuboi, H. Tatsuno, T. Nakashima, Y. Kajiyama, and Y. Nagasako, "Architecture and technologies of ATM subscriber network," *NTT R&D*, Vol. 42, No. 3, 1993, pp. 367–380.

Photonic Transport Network 4

This chapter describes photonic transport network technologies: the key to realizing future bandwidth-abundant transport networks. One important attribute that a multimedia network must possess is flexibility. Flexibility is best realized by the logical realization of various transport functions. The mechanism to enable this has been successfully developed—ATM technology, which was detailed in the previous chapter. Another important attribute that must be developed in the near future is cost-effective bandwidth-abundant physical transport capability. Broadband ISDN will fully penetrate our society only after these attributes are completely realized. In this chapter, key technologies to enable the second attribute are explored. We first identify future transport network requirements by reviewing the recent evolution of transmission and transport node technologies. The optical path technologies, which exploit WDM transmission and wavelength routing capability, are then discussed. It will be demonstrated how optical paths will enhance transport network performance. The important technologies to realize optical paths, optical path accommodation design, and optical path cross-connect hardware are then elaborated. The economics of the optical path network is also illustrated in this chapter.

4.1 PENETRATION OF PHOTONIC TRANSPORT TECHNOLOGIES INTO NETWORKS

4.1.1 Brief History of Optical Transmission Technology

In 1966, K. C. Kao and G.A. Hockham of Standard Telecommunication Laboratories predicted that low-loss optical fibers could be obtained by purifying raw materials and that optical fibers could provide a viable transmission medium; however, the fibers available at the time were not transparent enough to allow transmission over any significant distance [1]. In 1970, F. P. Kapron and others of Corning Glass Wool Co. succeeded in fabricating optical fibers with a transmission loss of 20 dB/km [2]. This success accelerated the subsequent development of low-loss optical fibers, resulting in a transmission loss of 0.2 dB/km [3] in 1979, which is almost the theoretical limit. In addition to low loss, fibers are superior to metallic cables in terms of their broad bandwidth, small diameter and weight, flexibility, and immunity to electromagnetic interference.

Together with optical fibers, the various optical components needed to realize optical-fiber transmission systems have been developed. In 1962, the pulsed-current oscillation of a semiconductor laser was attained for the first time [4]. Semiconductor laser technology progressed dramatically after the heterojunction laser was realized using an AlGaAs/GaAs structure [5]. This development reduced the threshold current density and improved radiance power levels, operating life, and linearity. Photodetectors, PIN photodiodes, and avalanche photodiodes became mature devices with enough performance for practical application by the end of the 1970s. This is true for other optical components, such as optical connectors and optical power couplers.

These advances in the 1970s allowed optical transmission systems to be implemented by replacing existing coaxial cable and pair cable transmission systems. Optical-fiber trunk transmission systems carrying 32-Mbps signals were first put into service in Japan in 1981 [6]. These systems use single-mode fibers with 1.3-μm zero-dispersion wavelength. Optical transmission systems have not only been applied to terrestrial transmission, but also to submarine transmission, and by the mid 1980s they were widely deployed.

4.1.2 Penetration of Optical Technologies into Networks

The penetration of optical technologies into the various network levels is schematically illustrated in Figure 4.1. As explained in the previous section, optical technologies were first applied to trunk transmission, because these technologies are best suited to this application, which requires long-distance and wide-bandwidth transmission. Moreover, the system cost is shared by many users (connections). The trunk transmission cost per bit is much lower with optical transmission than with metallic transmission. The opposite was true for subscriber access transmission until the middle of the 1990s because the degree of transmission facility sharing is much less in the subscriber application; in other words, facilities are much more dedicated to each subscriber. Furthermore, the bandwidth required for subscriber access is quite small compared to trunk transmission that multiplexes a significant number of channels. To attain cost-effectiveness in the subscriber application, facilities can be shared to some extent by accommodating several subscribers with one fiber by utilizing passive double star (PDS) or passive optical network (PON) techniques [7].

The optical technology applications addressed above realize only point-to-point transmission. However, to create large-bandwidth end-to-end connections

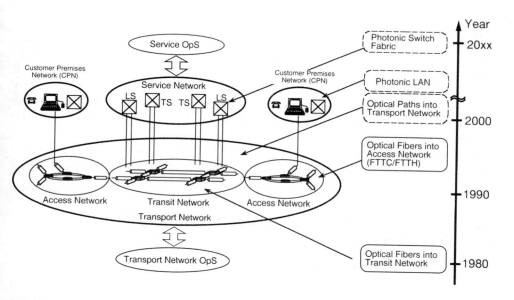

Figure 4.1 Penetration of photonic technologies.

cost-effectively, it is essential to reduce not only transmission cost but also transport node (cross-connect and ADM nodes) cost. The optical path technologies described in the following sections will play a key role in reducing both costs, and they are expected to be introduced in the network in the not-so-distant future.

The application of optical transmission technologies into local area networks (LANs) also started in the 1980s. Optical technologies have been utilized for fiber-optic Ethernet, fiber distributed data interface (FDDI) [8], and fiber channel [9,10]. FDDI uses multimode fibers with a preferred core diameter of 62.5 μm and a cladding diameter of 125 μm. The line rate is 125 Mbps using the 4B/5B binary coding scheme in which each 4-bit unit of information is mapped into a 5-bit word to facilitate physical-level transmission. Consequently, the maximum data rate at which a station can access FDDI is 100 Mbps. The fiber channel offers point-to-point connections at multiple data rates ranging from 100 Mbps to 800 Mbps. The transmission medium can be either single-mode fiber (up to 10 km), multimode fiber (up to 2 km), or coaxial cable (up to 50m). While high-speed LAN applications are possible with optical fibers, the penetration of optical-fiber transmission remains limited. This is because it is very cost-sensitive in this application area, and the transmission distances needed, which are relatively shorter than in other applications, are not always enough to justify the introduction. Photonic LANs, which utilize not only fiber transmission but also optical routing and other optical processing, if necessary, can be introduced widely only after optical processing technologies mature further and their cost is reduced to a level competitive with that of electrical technologies.

Photonic switching technologies have also been investigated widely [11–13]. They also need some time before they can be widely applied in the network to replace existing electrical switching facilities. This is because the technologies have not yet reached the level that allows them to be applied cost-effectively. Further advances in large-scale optical switching technology and optical memories are necessary before photonic switches can replace existing electrical switches on a large scale.

4.2 B-ISDN SERVICES AND TRANSPORT NETWORK REQUIREMENTS

4.2.1 Recent Advances in Transport Technology

Remarkable progress has been attained in this decade in the development of information transport network technologies. Optical-fiber transmission systems offering up to 2.5 Gbps have been introduced [14] very recently in Japan as well as those offering 10 Gbps [15,16]. SDH-based digital cross-connect systems (DCSs) and add-drop multiplexers (ADMs) have been or are going to be widely deployed, resulting in reduced transport network cost and reliability enhancement [17–19]. A breakthrough in transfer mode was seen with the development of ATM. ATM will be the base on which B-ISDN and multimedia communications should be built (Chapter 3).

On the basis of the transport technology advancement described above and in accordance with the increasing demands for data communication among business sectors, the frame relay service [20,21], switched multimegabit data service (SMDS) [22,23], and ATM network service [24,25] are being introduced. Their penetration is, however, limited to business subscribers, and no clear approach is currently known for the nationwide deployment of B-ISDN services, which should become as ubiquitous as the existing telephone service. The reason may be explained as the chicken or the egg conundrum; new services to stimulate high-speed demand are possible only if the end-to-end broadband transport capability is available at reasonable cost, which is possible only if demand is sufficient to allow network-wide transport platform development (see Figure 4.2). Although it may be true that a very attractive but bit-rate-intensive service exists but has not yet been identified, the key to this conundrum is the development of technologies that can create an open-ended communications network that offers large bandwidth at an affordable cost—several megabits per second—at twice the cost of the existing telephone service. The

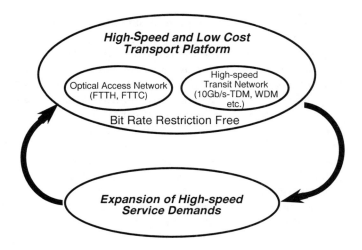

Figure 4.2 The chicken or the egg problem.

network will be the foundation on which the information society will be built and will stimulate attractive bit-rate-intensive services.

Subscriber networks have seen steps made toward the opticalization of subscriber loops, Fiber to the Home (FTTC) and Fiber to the Curb (FTTC), and this trend is expected to strengthen in the future [26–28]. Once optical fibers are introduced into subscriber networks, the bit-rate restriction of local access will become marginal and the transmission cost within the subscriber networks will be practically independent of the bit rate. In this context, trunk transport network throughput multiplication and cost reduction will be the keys to provide end-to-end broadband capability at low cost. Thus, the subsequent sections in this chapter focus on trunk transport networks.

4.2.2 Transmission Cost Requirements

One of the most likely bit-rate-intensive network services is video transmission. With utilizing recent advances in video coding techniques, MPEG I [29,30] and II [31] standardize the motion picture coding methods for digital storage media. The target transmission bit rate is less than 1.5 Mbps for MPEG I and less than 15 Mbps for MPEG II (main level). These bit rates are almost two orders of magnitude larger than that of digital voice transmission. No one will request the video transmission service if its price is two orders of magnitude higher than that of the existing voice transmission service. This is because video demands high bit rate; however, the effectiveness is not always proportional to the bit rate demanded.

Interesting research results were published in 1986 [32] regarding the effectiveness of voice and video communication. Ken'ichiro Hirota's comparison of the effectiveness of video and voice communication is summarized in Table 4.1. The first item is the relative measure of information transfer capability achieved by video and voice. It compares information quantity obtained by a two-dimensional color pattern (video) and that obtained by linguistic means. The result is that color video corresponds to 18,000 bits per minute while verbal language corresponds to 1,800 bits per minute; the ratio is 10 to 1. The second measure is the relative average cost of equipment and software. Comparing the ratio of equipment prices of color TVs to radios, a video recorder to a voice recorder, and VTR software to records, he found that the ratio is 6–12 to 1 (according to 1984 data). This means that if the equipment and the software costs of video are within that range, the video is widely accepted (people are prepared to pay 6–12 times more for video services). No subsequent investigations have been done recently on this item, and by virtue of the recent digital technology advances, the ratio seems to be decreasing now. The third is the information processing capability of a human being. If the communication channel is visual, humans can process the visual information at about 18,000 bits per minute while verbal information slows us down to 2,300 bits per minute; the ratio is 7.8 to 1. The last measure is the relative cost of advertising via either TV broadcasting or radio. The ratio was about 10 to 1 in Japan; in other words, video advertising is seen as 10 times more effective than radio

Table 4.1
Comparison of Video and Voice Communication

Item	Ratio of Video to Voice
Information transfer capability	10
Average related equipment and software cost	6–12
Information processing capability of human being	8
Cost for advertisement (on radio and TV)	10

From: [32], Ken'ichiro Hirota (Institute for Future Technology, Tokyo), 1986.

advertising. Including other comparison results, he concluded that, as a general consensus, video can be thought of as 10 times more effective than voice.

Of course, it is almost impossible to accurately discuss and evaluate the true relative effectiveness of video, since it depends on many factors including the services desired; however, the obtained value gives us a useful insight. As mentioned before, the bit rate required for video transmission (for example 6 Mbps of MPEG II) is roughly two orders of magnitude larger than that of voice/audio (for example, 64 Kbps). However, the effectiveness is less than one order of magnitude as explained above. These observations indicate that in order to attain wide acceptance of B-ISDN multimedia services, including video transmission, the infrastructure that supports them should transmit video signals with 100 times larger bandwidth than voice, at a cost of a few times that of voice transmission.

We will see the widespread deployment of broadband services including video transmission only after the network is realized where large bandwidth transmission is offered at very low cost.

4.3 TRANSPORT TECHNOLOGY EVOLUTION

Let us review the evolution of transport network technology in more detail [33] as depicted in Figure 4.3 using the layered transport model. The concept of a layered transport network architecture is important, as emphasized in Chapter 1. The following details are of Japan's network. Figure 4.4 shows the enhancements made to transmission capacity and cross-connect throughput.

4.3.1 Optical Transmission System Technology Advancement

Since the first demonstration of laser emission in 1960, tremendous efforts have been made to realize fiber-optic transmission systems. Remarkable progress has been seen, particularly in this decade. Notable developments include low-loss single-mode fibers and optical sources in the 1.5-μm wavelength region and ultra-high-speed electronic and photonic circuits [34]. The evolution of transmission capacity at NTT is depicted in Figure 4.4(a).

Optical-fiber transmission was first introduced in 1981, and since then the transmission capacity has been increased by more than one order of magnitude per decade [34,35]. This has resulted in a more than 90% reduction in cost in this decade. The continued enhancement of electronic devices has also driven research into high-speed transmission. Si-based high-speed electronics are now widely used in multi-Gbps transmission systems. Rapid progress in GaAs-based devices now permits operation speeds of several tens of gigabits per second. In fact, 2.5-Gbps transmission systems were introduced in Japan

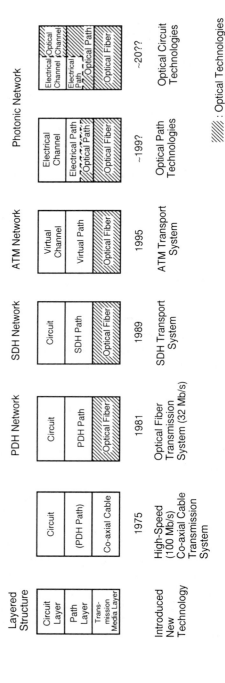

Figure 4.3 Transport network technology evolution in Japan. *After:* [33].

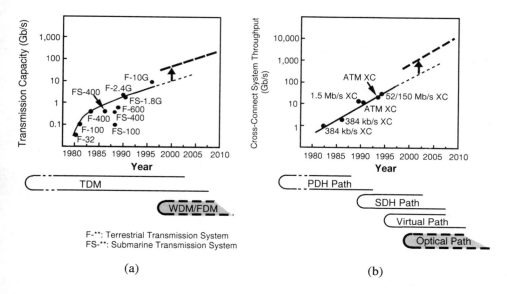

Figure 4.4 Transport network technology evolution: (a) transmission technology evolution and (b) path technology evolution (*After:* [33]).

in 1990, and 10-Gbps transmission systems are now being introduced [15,16]. Another technology that must be emphasized is the optical amplifier. In the last few years, significant progress has been recorded in the development of erbium doped fiber amplifiers (EDFAs) [36,37]. Their advantages include wide bandwidth, low noise, high gain/power, and easy connection to single-mode fibers. This technology greatly increases the application range of optical transmission by overcoming the optical loss of transmission, devices, and optical components. Furthermore, soliton transmission [38–40], dense wavelength division multiplexing (WDM)/optical frequency division multiplexing (FDM) [41–46], and high-speed optical signal processing [47,48] have the potential to enable further expansion of the transmission capacity and network flexibility.

The above developments open up new opportunities for provisioning broadband network services. Thus, the continually evolving optical-fiber transmission system technologies will form an important portion of the base on which the future bandwidth-abundant B-ISDN should be created.

4.3.2 Transport Node Technology Advancement

As for the path layer technology, cross-connect systems were first put into commercial use by NTT in 1980 for the PDH network, as shown in

Figure 4.4(b). They employ DS2 (6.3 Mbps) interfaces with line facilities, and the unit of cross-connection with time slot interchange techniques is 384 Kbps, or six telephone channels. (In 1981, DACS I was introduced by AT&T. It was designed to terminate up to 128 DS1 (1.5 Mbps) channels and the unit of cross-connection is DS0 (64 Kbps). It is used to support channel processing for special services and provides a convenient testing environment for T-1 carriers.) Standby path networks were implemented by NTT in 1982, with protection switches at the DS4 (100 Mbps) path level for a long-distance transmission network [49]. Protection switches at a lower path level (DS2) were introduced in 1985 [49]. The PDH networks of North America, Europe, and Japan employ different digital hierarchies, as seen in Chapter 2.

SDH paths (1.5 Mbps and 52 Mbps) were introduced [50] to Japan in 1989. The introduction of SDH has significantly simplified the existing plesiochronous digital hierarchy (PDH) path networks and allowed direct multiplexing/demultiplexing of paths so that the digital cross-connect function and transmission-signal-monitoring capabilities can be easily implemented (see Chapter 2). The digital cross-connect systems for 1.5-Mbps paths and those for 52-Mbps paths have been widely deployed. The 52-Mbps paths cross-connect systems are also utilized for protection switching [51].

The path is fundamentally a logical concept whose attributes are the bandwidth, route, and quality of service (QOS) that it provides. In SDH networks, a path is tightly linked to the physical interface structure for transmission. This linkage produces inefficiencies that can become a nuisance in the multimedia environment. For ATM networks, the virtual path (VP) (see Chapter 3) was standardized in ITU-T in 1990 [52]. The VP strategy enables the fully logical realization of path functions and can greatly enhance network capability through its flexibility. The VP benefits were elaborated in Section 3.3. The technical feasibility of a VP cross-connect system with a 1.4-Gbps interface and a throughput of more than 10 Gbps was experimentally confirmed [53] in 1990. The throughput can be increased to hundreds of gigabits per second in the near future [54,55]. Network restoration techniques with VPs are now being intensively studied [56–59]. VPs were introduced in 1994 for the pilot tests of multimedia services using a high-speed backbone network, ATM transport systems, and their management systems [60,61].

The throughput of cross-connect systems has been increasing in line with the introduction of SDH and ATM, as shown in Figure 4.4(b), and transport node cost (per bit) has been accordingly reduced. The ATM technology not only reduces transport node cost but also greatly enhances the flexibility of the path network, which will allow for the construction of multimedia networks.

One important point is that existing path layer approaches utilize electrical technologies (see Figure 4.3). In this chapter, a new technology, the optical path, is discussed.

4.3.3 Transport Network Restoration Technology Advancement

Network service robustness is a key issue for any network provider and its importance will keep growing. High-speed optical transmission systems are being extensively introduced and the social importance of data communication is increasing. Therefore, the severity of a single transmission system/link failure and its economic impact on society cannot be overestimated [62]. Failure immunization techniques and rapid restoration from failures are required. Different restoration techniques have been developed and utilized [63,64]. Physical layer restoration techniques such as automatic protection switching (APS), APS diverse protection, and dual homing have been implemented. These restoration techniques dedicate necessary facilities to restoration, so restoration time is generally very short, say less than 50 ms. On the other hand, they do require more spare facilities for attaining a specific restoration ratio than is needed with path layer restoration.

Path layer restoration is performed with digital cross-connect systems and add/drop multiplexers (ADMs) and does not necessarily require separate spare facilities. Instead, it uses the spare capacity of the working systems. Restoration flexibility is, therefore, superior to that possible with physical layer restoration. Path layer restoration can optimize the utilization of the available spare capacity for restoration, or the spare capacity needed can be minimized to attain a specific restoration ratio (for example, 100% restoration for any one link failure). These benefits are offset somewhat by the complexity of the control functions needed. Existing path restoration systems are all based upon centralized control.

These two kinds of techniques are generally combined to achieve maximum restoration performance. Figure 4.5 depicts the evolution of path restoration techniques in Japan's network. Protection switches at the DS4 (100 Mbps) path level were first implemented in 1982 for a PDH long-distance transmission network [49], as mentioned. Restoration at the VC-3 path level with digital cross-connect systems was introduced in 1995 [51] for SDH systems. ATM VP restoration systems are under development.

In addition to centralized control, recent technical advances in the field of distributed databases have accelerated the study of path restoration techniques with distributed control [65–67]. Furthermore, to match the replacement of STM with ATM, new techniques have been proposed to further strengthen network integrity [57–59] with VP restoration, as explained in Section 3.3.3.

The path restoration techniques developed thus far are specific to each path network, so some inefficiency is incurred. A new network failure restoration technique with optical paths is also explained in this chapter.

4.3.4 Next Steps to Realize Ubiquitous B-ISDN

As pointed out in Section 4.2, a tremendous transport network cost reduction is essential to enable broadband service explosion. To achieve this, quantum

128 Advances in Transport Network Technologies: Photonic Networks, ATM, and SDH

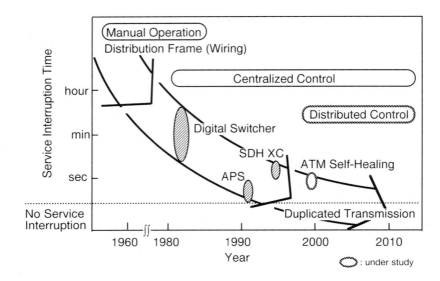

Figure 4.5 Evolution of path restoration techniques.

leaps in both transmission capacity and cross-connect throughput are needed simultaneously (see Figure 4.4, broken lines). ATM technology has revolutionized the signal transport mechanism, while its hardware can be realized with existing or reasonable enhancements of existing silicon-based electrical technologies. Accordingly, cross-connect system throughput is increasing in line with electrical technology advances. Relying on existing TDM technologies means that the transport bandwidth can grow only with further advances in semiconductor process technologies.

In the previous transport technology advances addressed herein, the point to be emphasized is that optical technologies have been introduced to just the physical media layer (see Figure 4.3); optical techniques have been utilized merely to increase transmission capacity. Expanding transmission capacity without a commensurate increase in cross-connect throughput will allow only inflexible and very limited networking of the transmission systems. This prevents the network-wide deployment of an end-to-end high-speed communication capability at low cost. In the next section, we will explore the key technology—WDM/FDM transport technologies—to solve these problems. Network integrity can also be attained effectively by employing WDM/FDM technologies, since network restoration within a WDM layer, which commonly supports different transfer mode networks, will be effective.

4.4 OPTICAL PATH TECHNOLOGY

4.4.1 Transport Network Photonization

The photonization of the transport network and the introduction of important transport techniques are depicted in Figure 4.3. As shown in Figure 4.3, optical technologies have been introduced to just the physical media layer. High-speed transmission systems such as 2.5-Gbps systems and 10-Gbps systems have already been introduced. These systems are based on TDM techniques, and hence optical techniques have been utilized merely to increase transmission capacity, as has been addressed. However, recent technical advances [41–46] in WDM/FDM techniques have reached a level at which their practical application seems feasible. When the available number of wavelengths (frequencies) on a single optical fiber is on the order of 10, the WDM (FDM) technique will be applicable to the path layer, and when it exceeds 10^2, the technique could be applied to the circuit layer. Here, an optical path [68–72] is basically a bundle of (electrical) circuits that is identified by its wavelength and routed at optical path cross-connects and ADMs according to its wavelength (wavelength routing). Considering the optical technologies that could be commercially available and cost-effective within the 1990s, this chapter highlights optical path technologies that utilize WDM techniques. (Most of the following descriptions regarding WDM are also applicable to FDM techniques; however, only WDM will be mentioned for simplicity.) The optical paths are used to accommodate electrical paths such as SDH paths and VPs, so a new optical path sublayer is introduced as depicted in Figure 4.3. Existing electrical paths such as PDH paths, SDH paths, and VPs could converge into a single VP layer in the future multimedia environment. However, no plausible technical forecast has indicated that optical paths will replace all electrical paths within 10 to 20 years. In other words, optical paths will complement electrical paths, replacing a portion of the electrical path functions and enhancing the performance of the path layer. How optical path technologies will enhance network potential is explained in the following section.

4.4.2 Benefits of WDM

The application of optical technologies to the path layer will offer significantly more benefits than the simple enhancement of transmission bandwidth. WDM technologies enable this because they can significantly expand not only transmission capacity, but also cross-connect node throughput by capitalizing on the wavelength routing capability [73-76] of paths. In this respect, WDM offers more benefits than TDM or space division multiplexing (SDM). Other potential WDM benefits over TDM are listed below.

- With WDM, only that portion of the total line capacity that needs to be dropped at a node can be terminated. TDM, on the other hand, terminates the entire line capacity when accessing the transmission line. This is a remarkable benefit when the total line capacity is very large, since it minimizes the use of ultra-high-speed electronics technology, which is characterized by large power consumption and the need for serial to parallel conversion of the bit stream.
- The total throughput of an optical path cross-connect system (see Section 4.6) can be much larger than that of an electrical TDM cross-connect system, and the hardware can be simplified (as will be demonstrated in Section 4.6), since no synchronization is needed among optical paths. This nonsynchronous characteristic enables us to devise a cross-connect system with highly modular growth capability, as will be discussed in Section 4.6.

Accordingly, WDM technologies will be exploited not only to enhance transmission capacity without constructing outdoor facilities (additional fibers and ducts), but also to expand cross-connect node throughput by capitalizing on wavelength routing.

4.4.3 Benefits of Optical Path Technologies

This section overviews the benefits of optical path technologies. The details, including quantitative evaluations, are discussed in the following sections.

Enhanced Transmission Capacity

Both 156-Mbps and 622-Mbps user-network interfaces (UNIs) have been standardized in ITU-T [77] and the 156-Mbps subscriber access will soon be commercially available. When optical technologies are applied to the subscriber network, local communication will be free from bit rate restrictions. However, trunk transmission capability is rather limited. For example, one 10-Mbps long-haul channel can be provided to each existing customer only at an excessively high transmission cost. (The target is just a few times the cost of a conventional telephone channel (64 Kbps), as discussed in Section 4.2.) Transmission capacity multiplication with low cost is therefore necessary to deploy a cost-effective B-ISDN and to enable a network paradigm shift to a true broadband ISDN. Optical path technologies utilizing WDM can provide a viable means to achieve the full potential of optical-fiber transmission: several terabits per second [46].

Enhanced Cross-Connect Node Throughput

In conjunction with the transmission capacity increase made possible by optical technologies, the cross-connect node throughput must also be expanded to create an effective network. When optical path technologies are utilized, this is a natural consequence because the electrical processing bottleneck is eliminated through the introduction of wavelength routing [73–76] of paths at cross-connect nodes. The total throughput of an optical path cross-connect system can be much larger than that of an electrical TDM cross-connect system, and the hardware can be simplified (as will be explained in Section 4.6), since no synchronization is needed among optical paths. Figure 4.6 depicts optical path cross-connect throughput in terms of input/outgoing port number, bandwidth of a path, and number of paths(wavelengths) multiplexed within a fiber.

Flexible Service Provisioning

At the outset of B-ISDN introduction, B-ISDN demands may emerge abruptly over a large geographical area, and although the required number of B-ISDN connections may be small, the aggregate bandwidth can be very large compared to existing narrowband services. The optical path layer allows new path connections to be provided on demand without the construction of additional outdoor equipment, such as ducts or new cables, and thus enables service provisioning with minimum delay. This produces a significant practical benefit, especially in long-haul transmission, given the recent environment of high outdoor facility construction cost in urban areas.

With WDM, as mentioned above, modular growth capability of the cross-connect system can be more easily devised, as will be described in Section 4.6. This is a very important point for the economical introduction of a cross-connect system with the potential for vast throughput to transport nodes whose needed throughput ranges from small to large.

Thus, the flexible and cost-effective broadband service provisioning capability and the transport expansion capability provided by optical paths will play important roles in the next stage of transport network evolution.

Economical Broadband Transport Network Construction

Figure 4.7 explains one of the major benefits of optical paths. If WDM technologies are applied to the physical media layer or to enhance transmission capacity, as shown in Figure 4.7(b), transmission cost (per bit) can be reduced because no construction of outdoor facilities is needed; this effect is almost the same when TDM is applied instead of WDM. When WDM technologies are also

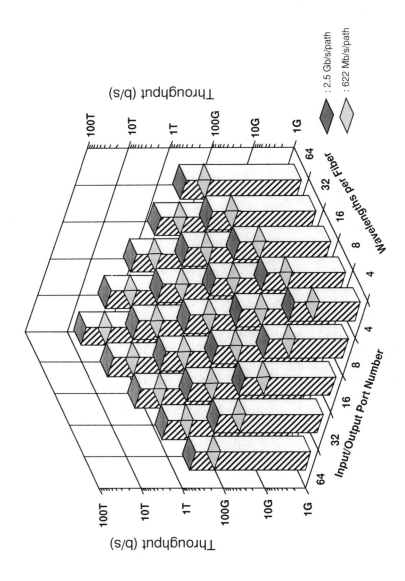

Figure 4.6 Optical path cross-connect system throughput.

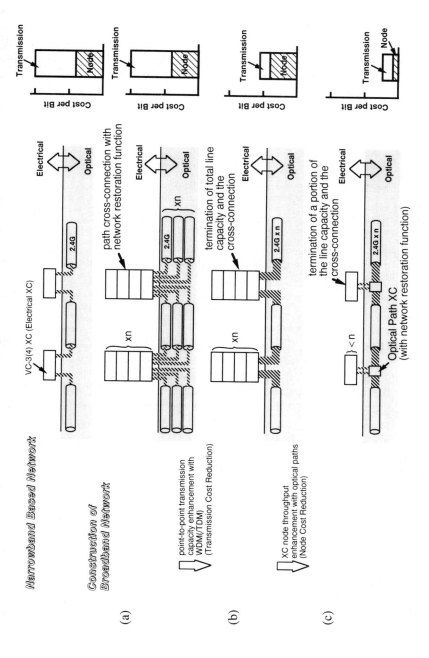

Figure 4.7 Transport network cost reduction with optical paths: (a) broadband network with existing electrical system, (b) broadband network with WDM for transmission media layer, and (c) broadband network with WDM for path layer. (Source: [154]. © 1994 IEEE.)

applied to the path layer, as shown in Figure 4.7(c), optical paths can be cross-connected and the network restoration function can be performed within optical paths, as described later in this section. At a transport node, all traffic except that terminating at the node is cross-connected at the optical level with wavelength routing, which eliminates the electrical processing bottleneck so node cost reductions can be achieved. The cost reduction possible with optical paths is discussed in Section 4.7.

Optical Platform

Transmission technologies are evolving from PDH and SDH to ATM. Considering ATM, in addition to the already standardized cell transmission format based on the SDH frame, the format based on a continuous stream of cells (called the *cell-based ATM transport network*) may be fully standardized in the future [78]. The optical path layer imposes basically no restriction on the electrical path transmission mode carried by the optical paths; even analog signals are possible if desired. Thus, the optical path layer can provide a platform that can subsume different transfer-mode networks. As for the application of optical paths, however, restricting transmission signal formats according to the application areas may allow more cost-effective networks [79].

Existing networks must work with different transmission modes, and, consequently, transmission-mode conversion systems are required when two different mode signals are transmitted within the same optical fiber (see Figure 4.8). To this end, a transmission format converter (adapter (ADP) in Figure 4.8) linking two SDH VC-3 higher order paths and one PDH 100-Mbps level path was developed [80]. An STM-ATM conversion system (STM/ATM Conv. in Figure 4.8) has also been developed [81]. On the other hand, if optical paths are utilized, different transmission modes can be carried by different optical paths, if so desired. Such optical paths are accommodated within the same optical fiber and use the same optical path cross-connect system. This is schematically illustrated in Figure 4.8. How the network layer architecture, including the optical path layer, should be determined, or what kinds of transmission modes should be supported with the optical path layer, are important subjects to be determined. For example, if the number of available wavelengths per fiber is relatively small, say less than 30, optical paths supporting SDH (and SDH-based ATM signals) will be efficient in creating nationwide backbone networks [82,79]. However, optical paths supporting analog subcarrier multiplexing (SCM) signals may be effective for local access networks wherein video distribution services can be of great importance.

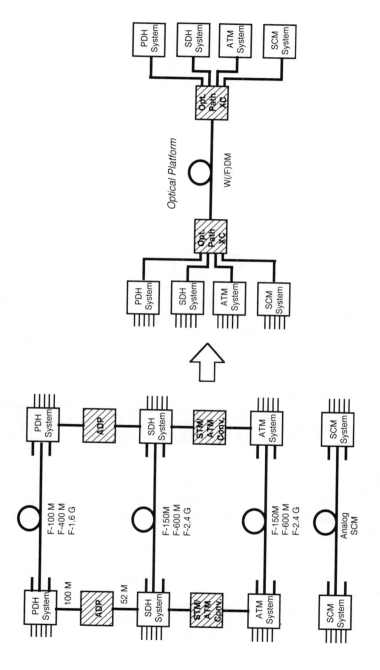

Figure 4.8 Encapsulation of different transport modes with optical paths (*After:* [72]).

Failure Restoration With Optical Paths

The encapsulation capability of different transmission modes can be exploited to develop an effective network restoration mechanism. As described in Section 4.3.3, different network restoration techniques have been implemented for each transmission-mode network. If the optical path layer is utilized for network restoration, a major portion of the network restoration systems would be used in common among different transmission-mode networks, while failure detection and notification schemes would be unique to each network. This would yield more effective utilization of the network spare capacity reserved for restoration since the spare capacity at the optical path level is shared by all transmission-mode networks supported by the optical paths. Of course, this is only possible when optical paths are defined to support all transmission-mode networks.

The optical path layer restoration is particularly effective for ATM networks in which some adaptation may be required between VPs (that can have fine granularity) and the restoration VP unit (virtual path group (VPG) [83]). The reason is explained in the following.

Restoration time is a critical issue in B-ISDN and must be minimized. An evaluation example of the effect of the number of restoration paths on the restoration time is presented below using an existing network model. In existing path restoration systems, the expected maximum number of restoration paths accommodated within a transmission line is 16 (1.6 Gbps/100 Mbps) for a PDH network and 48 (2.4 Gbps/50 Mbps) for an SDH network. On the other hand, in an ATM network it is 4,096 (12 VPI bits), as shown in Figure 4.9. In general, the smaller the average path capacity is (the finer the restoration granularity is), the smaller is the required spare network capacity reserved for restoration or the larger the average link utilization efficiency is, considering network restoration. On the other hand, the larger the number of paths to be restored, the longer the restoration route search time becomes.

The required computing time was evaluated [72] for finding restoration path routes for one transmission route (a bundle of optical fibers connecting two nodes) failure considering Japan's nationwide long-haul transmission network model, which consists of 260 transmission nodes and 800 links. The evaluation assumes a centralized control scheme for restoration. Two restoration schemes, line restoration and path restoration, were tested. In the line restoration scheme, restoration paths were allocated between a pair of nodes, terminating the failed line. In the path restoration, restoration paths are allocated between the terminating nodes of each failed path.

Other assumptions adopted are as follows.

- Traffic volume (number of active paths) between any two nodes is randomly set.

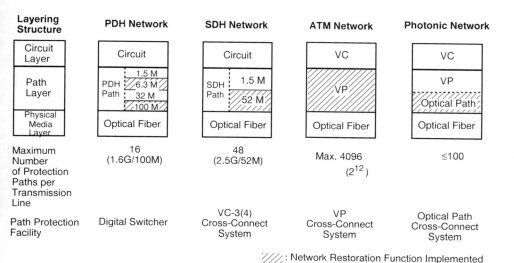

Figure 4.9 Allocation of network restoration function (*After:* [72]).

- A cross-connect system (restoration switch) is installed in each node.
- All paths traversing or terminated at a node are accommodated within a cross-connect system.

The restoration path route search algorithm used [72] was a slightly modified version of Dijkstra's algorithm. The calculation was terminated when the minimum length path route between the two nodes was determined. This algorithm may not be an optimum one in terms of required spare capacity for restoration paths, however, it does provide relatively short restoration route search times. In the simulation, the mean calculation time needed to determine restoration path routes was evaluated for single route failure on all routes in the network. The simulation used C language programs running on a 1.6-MFLOPS workstation (SUN 4).

Figure 4.10 shows the evaluation results for line and path restoration where the average number of paths per each fiber (traffic volume) was altered. It is shown that the required restoration route calculation time increases almost linearly with the number of paths per fiber. Actual restoration time should include database search time and restoration message transmission and message processing time in addition to the evaluated restoration route calculation time. Considering this, and the expected available computing power, it is concluded that the maximum number of paths per fiber should be less than the order of

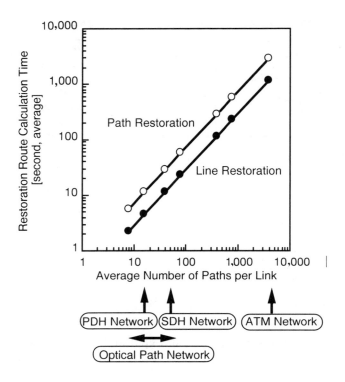

Figure 4.10 Restoration path pattern calculation time (*After:* [72]).

10^2 for this scale of network, assuming the restoration time target is to be within several seconds. Therefore, in ATM networks, grouping of VPs is necessary for VP restoration [83] in order to reduce the number of restoration paths per fiber. This, however, entails the definition of a new architectural level of VPG, as mentioned before, and also requires the development of mechanisms needed to perform operation, administration, and maintenance (OA&M) for the level. In this context, optical path layer restoration will be especially effective in ATM networks. The envisaged granularity of the expected WDM optical paths (maximum number of wavelengths multiplexed within a fiber) falls within the desirable range, as shown in Figure 4.10.

The technical details of optical path layer restoration will be explained in Section 4.5 in conjunction with optical path accommodation design issues within a physical network.

4.4.4 Optical Path Realization Technologies

Various lightwave network technologies have been proposed. In this section, we focus on optical paths accommodating electrical paths and clarify the suitability of each lightwave network technology for the optical path layer. Optical path realization techniques are classified in Figure 4.11. Optical paths can be divided into two categories. One category includes the optical paths transporting electrical signals using a cell/packet format. Such paths are called ATM optical paths herein. The other paths are named wavelength paths/virtual wavelength paths (WPs/VWPs) [68–72], and can basically support all electrical transfer modes such as STM and ATM.

ATM Optical Paths

ATM optical paths can be divided into single-hop or multihop paths in terms of the cell routing scheme on the electrical path (VP) to the destination node. This is explained in terms of the hierarchical level-to-level relationship shown in Figure 4.12 where connection endpoints for the VPC and optical path connection (OPC) are the same in a single-hop network. VPCs may traverse multiple OPCs, as mentioned before, where connection endpoints for the VPC and concatenated OPCs are the same in a multihop network. In a single-hop network, cells are directly transported to the electrical path destination node on an optical path without requiring electrical conversion in between. The simplest physical topology to enable this is a star topology [84,85], which can be realized using an optical star-coupler in the hub node, as shown in Figure 4.13.

In this network, all nodes are connected together with a star coupler. Two communication strategies are utilized; a wavelength is assigned to the sending node or to the destination node. In the former case, in the communication phase, each node broadcasts cells (packets) to all other nodes using a specific wavelength assigned to each sending node (fixed-wavelength transmitters), and the destination node receives the specific wavelength cells using a wavelength demultiplexer or a tunable-wavelength filter, or equivalent functional devices (tunable receivers). In the latter case, each node broadcasts cells at a specific wavelength to the destination node using a tunable-wavelength optical transmitter. The destination node selects specific wavelength cells using a fixed-wavelength receiver. Thus, the network utilizes a broadcast-and-select strategy.

Signaling may be employed to establish connection before the communication phase, or collision detection at the receiver nodes and a contention resolution scheme may be employed. The signaling can be performed with the same star network or a different network developed specifically for signaling. With this scheme, one-to-many communication is easily performed since the broadcasting strategy is utilized. Some of these single-hop network examples and characteristics of each network are shown in Table 4.2.

A single-hop network thus leads to cell collision/contention, which demands the dynamic cooperation of nodes [84,85]. Therefore, the applicability of the single-hop network is limited to small physical scale networks such as

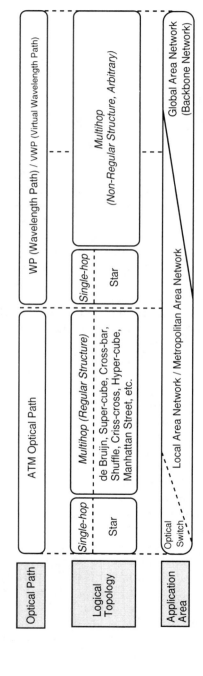

Figure 4.11 Classification of optical path realization technologies (*After:* [72]).

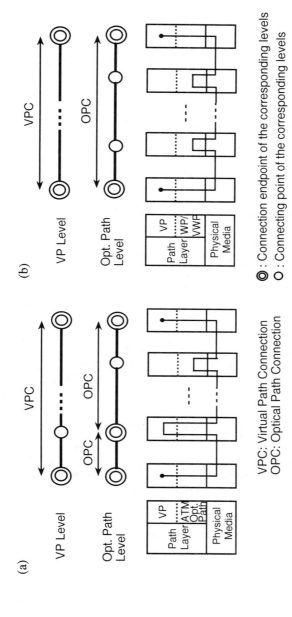

Figure 4.12 Comparison of (a) multihop and (b) single-hop ATM optical paths—hierarchical level-to-level relationship.

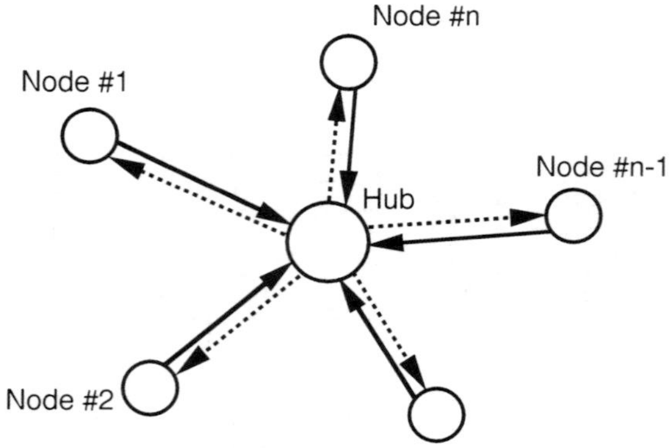

Figure 4.13 Star-coupler-based single-hop network.

LANs/MANs. This configuration is also being considered for application to the optical switch fabric [42].

On the other hand, in a multihop network, cells on an electrical path are transported to the destination node through concatenated multiple optical paths, determined with a regular structure (multihop), and undergo electrical conversion at each optical path endpoint so that routing of cells (selection of the next optical path to be traversed) can be done according to the electrical cell header information.

Multihop networks utilize various regularly structured logical connections [86–90] such as perfect shuffle [91–96], hypercube [97], cross-bar [98], Manhattan Street [99,100], criss-cross [98], supercube [101], and de Bruijn Graph [102,103], and wavelength assignment is performed for logical connections. Some of the logical connections are depicted in Figure 4.14. Each logical connection link connecting two nodes corresponds to an ATM optical path where a wavelength is allocated according to the rule determined by each logical connection, as depicted in Figure 4.15. Thus, mapping a logical connection into a physical transmission line network is required [99,104,105]. In this process, however, problems related to wavelength assignment are generally simpler than those in the WP scheme. We explain this procedure in Figure 4.15, which shows how a VP network can be accommodated within a physical network via multihop ATM optical paths.

WP and VWP

In the wavelength path (WP) scheme, each optical path is established between two nodes as required by allocating one wavelength for the path, as shown in

Photonic Transport Network 143

Table 4.2
Examples of Single-Hop Point to Multipoint Network

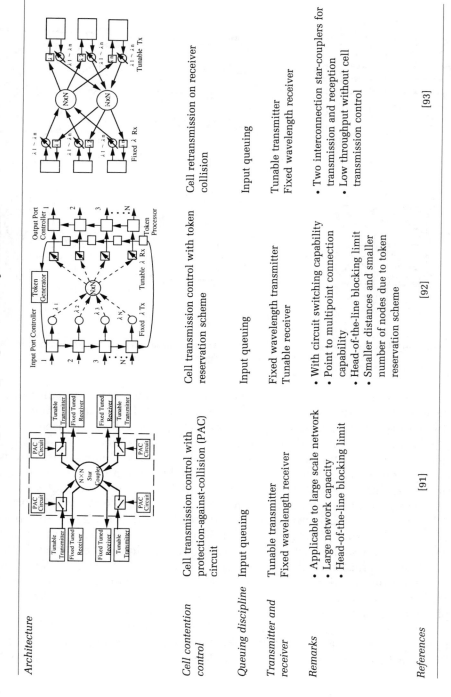

Architecture			
Cell contention control	Cell transmission control with protection-against-collision (PAC) circuit	Cell transmission control with token reservation scheme	Cell retransmission on receiver collision
Queuing discipline	Input queuing	Input queuing	Input queuing
Transmitter and receiver	Tunable transmitter Fixed wavelength receiver	Fixed wavelength transmitter Tunable receiver	Tunable transmitter Fixed wavelength receiver
Remarks	• Applicable to large scale network • Large network capacity • Head-of-the-line blocking limit	• With circuit switching capability • Point to multipoint connection capability • Head-of-the-line blocking limit • Smaller distances and smaller number of nodes due to token reservation scheme	• Two interconnection star-couplers for transmission and reception • Low throughput without cell transmission control
References	[91]	[92]	[93]

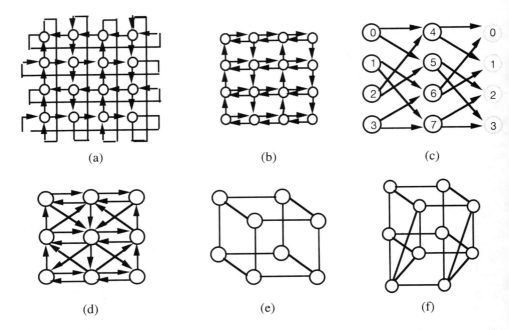

Figure 4.14 Regular logical network topologies: (a) Manhattan street, (b) cross-bar, (c) perfect shuffle, (d) criss-cross, (e) hypercube, and (f) supercube.

Figure 4.16(a). The intermediate nodes (WP cross-connect nodes) along the WP perform WP routing according to the wavelength (wavelength conversion is not performed within WP cross-connect nodes). The WP can be used, for example, for the internode path, which accommodates a number of VPs between the node systems (switching units, for example). Even in this case, if there is not enough traffic (enough VPs with a certain total bandwidth) between some pairs of nodes, then direct WPs between these pairs of nodes will not be set. The VPs between these pairs of nodes must traverse multiple WPs. Thus, multiple WPs are needed to accommodate a VP if no direct WP is available. In this sense, WPs (VWPs) can be thought of as multihop networks. WPs are accommodated within a physical network, and, in the accommodation process, the wavelength assignment problem must be solved simultaneously so that different wavelengths are assigned to each WP in a link fiber throughout the network [106,107]. The relationship among an electrical path (VP) network, a WP network, and a physical media network is also explained in Figure 4.15.

On the other hand, in the VWP scheme the VWP wavelength is allocated fiber by fiber (see Figure 4.16(b)) and thus the wavelength of each VWP on a fiber has local significance instead of global significance as in the case of WPs.

Photonic Transport Network 145

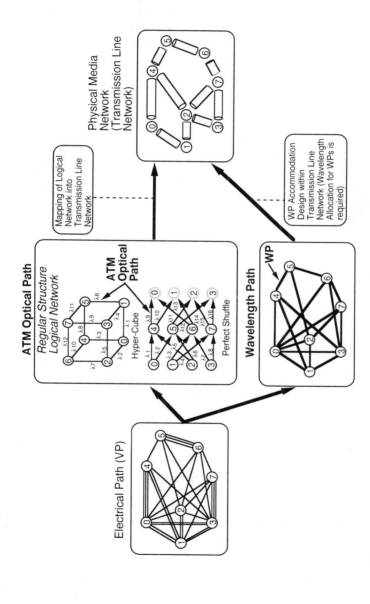

Figure 4.15 ATM optical path (multihop) and wavelength path (*After:* [72]).

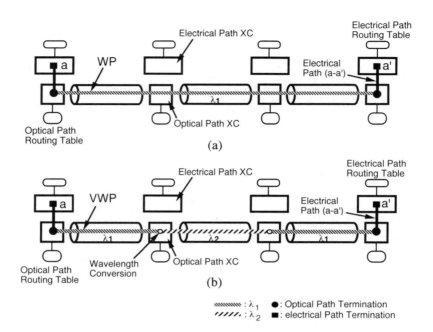

Figure 4.16 (a) WP and (b) VWP schemes. (*Source:* [154]. © 1994 IEEE.)

This is similar to the virtual path identifier (VPI) assignment principle in ATM networks. For this reason, this scheme is called virtual WP. Note that VWP termination is done at both originating and destination nodes and the intermediate nodes perform VWP routing (cross-connection) according to a routing table, along with any necessary wavelength conversion. Comparisons of WPs and VWPs are discussed later in this section.

By imposing restrictions on the physical configuration of a network, the WP network can be simplified. Some of the examples are given below. One example is to utilize a single-hop physical star network, as shown in Table 4.3. In other words, simple wavelength routing is achieved by utilizing a physical star network; however, the network capability is very limited. To create a physical star connection, an optical star coupler or wavelength multiplexers and demultiplexers are utilized in the hub node. Examples of the former scheme are LAMBDANET [108,109] and the broadband WDM star network (see Table 4.3). Examples of the latter are given in [110,111] (see Table 4.3).

With an optical star coupler, the broadcasting scheme (point-to-multipoint connection) is utilized and the network capability is determined by the functional performance of the end nodes. In each LAMBDANET node, one wavelength-tuned transmitter and $n - 1$ wavelength-tuned receivers (and another

Photonic Transport Network 147

Table 4.3
Examples of Single-Hop Star Network

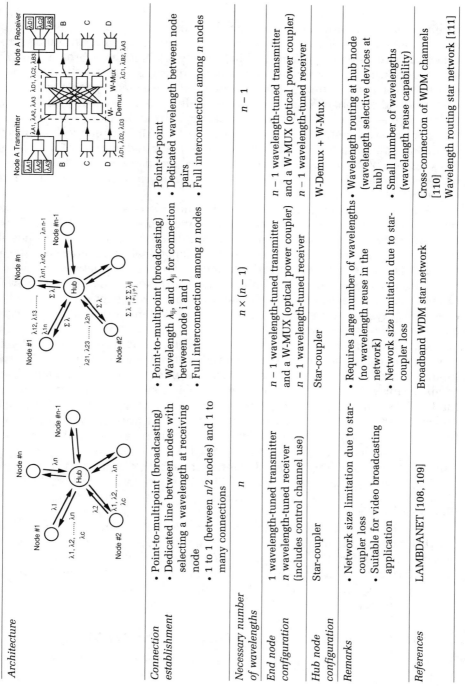

Architecture			
Connection establishment	• Point-to-multipoint (broadcasting) • Dedicated line between nodes with selecting a wavelength at receiving node • 1 to 1 (between n/2 nodes) and 1 to many connections	• Point-to-multipoint (broadcasting) • Wavelength λ_{ij} and λ_{ji} for connection between node i and j • Full interconnection among n nodes	• Point-to-point • Dedicated wavelength between node pairs • Full interconnection among n nodes
Necessary number of wavelengths	n	$n \times (n-1)$	$n-1$
End node configuration	1 wavelength-tuned transmitter n wavelength-tuned receiver (includes control channel use)	$n-1$ wavelength-tuned transmitter and a W-MUX (optical power coupler) $n-1$ wavelength-tuned receiver	$n-1$ wavelength-tuned transmitter and a W-MUX (optical power coupler) $n-1$ wavelength-tuned receiver
Hub node configuration	Star-coupler	Star-coupler	W-Demux + W-Mux
Remarks	• Network size limitation due to star-coupler loss • Suitable for video broadcasting application	• Requires large number of wavelengths (no wavelength reuse in the network) • Network size limitation due to star-coupler loss	• Wavelength routing at hub node (wavelength selective devices at hub) • Small number of wavelengths (wavelength reuse capability)
References	LAMBDANET [108, 109]	Broadband WDM star network	Cross-connection of WDM channels [110] Wavelength routing star network [111]

receiver for the control channel) are used for information transfer, which allows 1 to 1 connection between $n/2$ (when n is even) node pairs, simultaneously. The number of wavelengths needed in the network is n. In the broadband WDM star network, $n - 1$ wavelength-tuned transmitters, a wavelength multiplexer, and $n - 1$ wavelength-tuned receivers are used in each end node, which allows each node to establish connections to the other $n - 1$ nodes at the same time, thus $n \cdot (n - 1)$ connections in the network, simultaneously. The necessary number of wavelengths in the network is $n \cdot (n - 1)$. This network allows full interconnections among n nodes; however, the required number of wavelengths is excessive when n becomes large. This is because wavelengths cannot be reused in the network (i.e., each connection in the network utilizes a different wavelength).

On the other hand, with wavelength multiplexers and demultiplexers at the hub node, full-mesh point-to-point connections can be established with $n - 1$ wavelengths as shown in Table 4.3 (right-hand side). This is made possible by employing wavelength routing with wavelength-selective devices at the hub node. This configuration permits wavelength reuse in the network.

The networks shown in Table 4.3 are simple in terms of their physical topology and connection establishment principle (wavelength allocation); however, the physical topology restricts network expandability and application extent to rather small networks.

Another simple WP network can be developed by restricting the physical topology to a chain or a ring, as shown in Table 4.4. By restricting the physical topology, each node is composed of only wavelength multiplexers and demultiplexers for wavelength routing; space division switches are not required. If the WP/VWP networks are to be free of any physical topology restrictions, the network nodes require space division switches or their equivalents (combination of star couplers and wavelength-tunable filters, for example), as will be explained in Section 4.6. For these networks, the number of wavelengths required for establishing full-mesh connections (both directions) is on the order of n^2, and the ring network requires about twice as many wavelengths as the chain network, while the chain network needs two fibers for setting up connections in both directions [111].

Specific physical topologies such as the star, chain, and ring simplify network node configuration and the routing principle; however, the topology requirement hinders the development of a network that is large-scale and flexible in terms of network expandability and adaptability to traffic demand increases. Therefore, the following sections will discuss WP/VWP networks that impose no such a network topology constraint.

Comparisons of ATM Optical Paths (Multihop) and WPs/VWPs

ATM optical paths (multihop) and WPs are compared in Table 4.5. As shown in Table 4.5, in the multihop ATM optical path scheme each node has a small

Table 4.4
Wavelength Routing Networks With a Specific Physical Topology

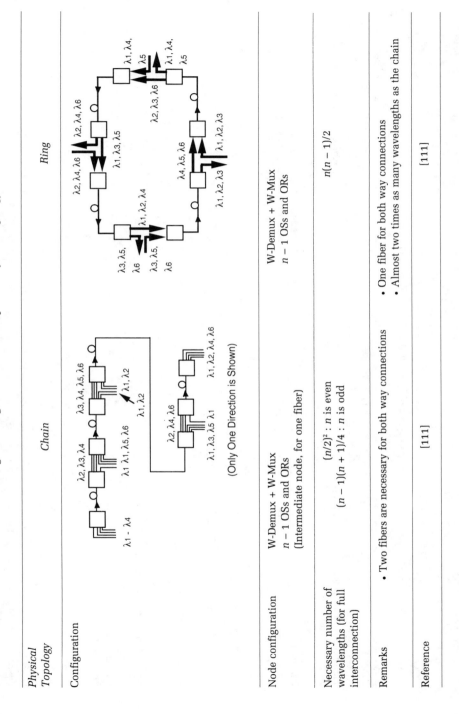

Physical Topology	Chain	Ring
Configuration	(see figure)	(see figure)
Node configuration	W-Demux + W-Mux $n-1$ OSs and ORs (Intermediate node, for one fiber)	W-Demux + W-Mux $n-1$ OSs and ORs
Necessary number of wavelengths (for full interconnection)	$(n/2)^2$: n is even $(n-1)(n+1)/4$: n is odd	$n(n-1)/2$
Remarks	• Two fibers are necessary for both way connections	• One fiber for both way connections • Almost two times as many wavelengths as the chain
Reference	[111]	[111]

Table 4.5
Comparison of ATM Optical Path and Wavelength Path

Items	ATM Optical Path (Multihop)	Wavelength Path (/Virtual Wavelength Path)
Transmission format on electrical path level	Cell-Packet	Basically no restriction (depends on interface specification)
Cell routing between nodes	Wavelength routing (regular logical topology) + electrical ATM cross-connect	Wavelength routing (arbitrary logical topology) + (electrical SDH/ATM cross-connect; when direct WP/VWP does not exist)
Total throughput	Smaller	Larger
Optical interface cost	Smaller (small number of OS/OR per node)	Larger
Required number of wavelengths in the network	Smaller	Larger
Network resource utilization	Lower	Higher
Electrical path level processing	More	Less
Transport delay	Larger (multihop via several nodes that may be located a long way away)	Smaller (Multihop can be minimized)
Optical path accommodation design within physical media network	Mapping of adopted logical topologies into physical layer topology	Path accommodation design (with wavelength assignment for WP)

number of optical transmitters and receivers [94]. This significantly reduces optical interface cost, a salient benefit of the scheme. However, with the multihop ATM optical path scheme, a great percentage of electrical cells hop multiple optical paths according to the logical network adopted, with electrical processing at each optical path end node (electrical cell routing is based on store-and-forward process). Consequently, the total transmission delay tends to be so large that the applicability of the scheme in terms of network scale is rather limited. The scheme is less suited to public telecommunication infrastructure than the WP scheme. Other features are compared and discussed in [72]. The envisaged application area for each optical path technique is also shown in Figure 4.11. The optical path technologies that are most suited to global area networks, therefore, are WP and VWP, which will be discussed in detail hereafter.

Comparisons of WPs and VWPs

The WP scheme described above holds various advantages; however, the WP accommodation design within a physical network is much more troublesome than that for digital paths (PDH or SDH paths) [112,113] and VPs [114]. This is because, as mentioned before, in determining the optical path route, optimization is necessary not only for total path length but also for the required number of fibers in a network under the conditions that the maximum number of WPs that can be accommodated within a fiber is limited and that wavelength assignment must be achieved such that all WP wavelengths on the same fiber are different. With the WP scheme, therefore, wavelength reuse in the network is limited to some degree. In large-scale networks, the WP accommodation design imposes a considerable processing burden. This tends to make the calculation time needed to find restoration path routes impractically long for real-time restoration path route searching in the case of network failure. Several heuristic methods have been proposed for WP accommodation design that are applicable to large-scale network design, which will be described in Section 4.5.

Figure 4.17 shows a comparison of WP and VWP schemes and the path connect tables at path cross-connect nodes. As shown Figure 4.17, to establish four WPs (Figure 4.17(a)) and four VWPs (Figure 4.17(b)) with the same routes, the WP scheme requires three wavelengths. On the other hand, the VWP scheme requires only two. Generally speaking, the VWP scheme allows the degree of wavelength reuse in the network to be maximized so that fewer wavelengths or fewer fibers are needed than with the WP scheme. The extent of the wavelength (or fiber) reduction depends on various conditions. The major ones are physical network topology, number of paths between nodes, and restoration path

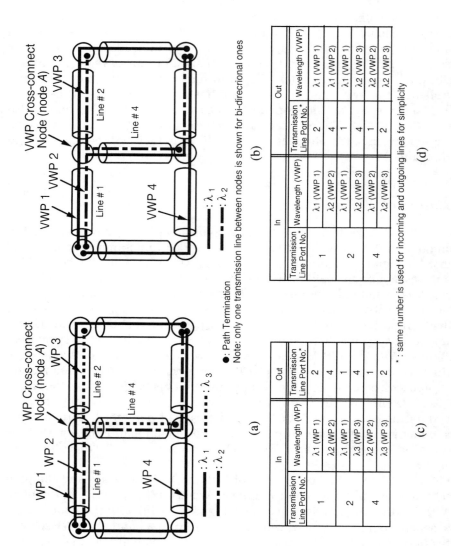

Figure 4.17 Comparison of WP and VWP: (a) WP, (b) VWP, (c) WP routing table example in node A, and (d) VWP routing table example in node A.

arrangement principles (line or path restoration). Quantitative comparisons are given in Section 4.5.

Another significant benefit of VWPs is the resultant simple VWP accommodation design (i.e., there is no wavelength allocation problem). In VWP accommodation design, the only condition is the maximum number of VWPs (or wavelengths) that can be accommodated within a fiber. This condition is basically the same as that for digital paths or VP accommodation design (for digital paths, the condition is the maximum number of paths within a fiber, and for VP accommodation, the total bandwidth of VPs accommodated in a fiber). Therefore, the restoration path route calculation time shown in Figure 4.10 is applicable to the VWP scheme, but not to the WP scheme. The WP and VWP schemes are compared in Table 4.6. Other major advantages of the VWP scheme are as follows:

- Expandability of the network (links and nodes) is larger since no restriction is imposed in terms of wavelength assignment in the network.
- Relative preciseness of wavelengths is required only between link-terminating nodes since utilized wavelengths are independent link by

Table 4.6
Comparison of WP and VWP

Item	Wavelength Path	Virtual Wavelength Path
Wavelength conversion at optical path XC	Unnecessary	Necessary
Wavelength assignment problem	Exist	Does not exist
Restriction of network and path expansion	Significant	Insignificant
Required wavelength number in the network (routing freedom)	Larger (smaller)	Smaller (larger)
Path wavelength allocation/control	Requires centralized control in the network	Possible with link-by-link distributed control
Wavelength precision	Absolute precision throughout the network	Link-by-link relative precision

After: [72].

link. The WP scheme demands absolute preciseness of the wavelengths throughout the network.

- Since each VWP cross-connect has a wavelength conversion function, different types of single-mode fibers such as 1.3-μm single-mode fibers and 1.5-μm dispersion-shifted single-mode fibers can be used to accommodate each VWP if VWP cross-connect nodes are designed to enable this (see Figure 4.18). This is because no VWP termination at the intermediate node is necessary even when the incoming and outgoing fiber types are different (wavelength conversion is necessary). On the other hand, in the WP scheme, the WP must be terminated when the incoming and available outgoing fiber types are different and cross-connection at the electrical path level is required, as shown in Figure 4.18(a). This VWP characteristic may enhance the extent of its application.

The most significant VWP weakness is the wavelength conversion function needed at cross-connects. This requirement offsets its known superiorities. The choice, therefore, depends on technologies available and cost. The feasibility of this function determines the effectiveness of the VWP scheme. The hardware realization of WP and VWP cross-connects is discussed in Section 4.6.

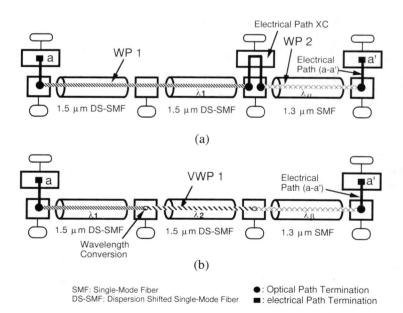

Figure 4.18 (a) WP schemes and (b) VWP schemes with different wavelength region fibers. (*Source:* [154]. © 1994 IEEE.)

4.5 WP AND VWP ACCOMMODATION DESIGN

4.5.1 Path Accommodation Design Without Considering Network Restoration

To develop optical path networks, optical path accommodation design must be developed within the constraints of the transmission media network. As mentioned in Section 4.4.4, for VWPs, it is basically the same as the electrical path accommodation problem, while for WPs, the wavelength assignment problem must be solved simultaneously so that no WPs sharing any fiber are allocated the same wavelength.

Different conditions and objective functions with regard to the WP establishment problem have to be considered depending on the application. One is a constraint on the number of wavelengths that can be multiplexed in a fiber. In the future, this constraint will be mitigated with further advances in optical technologies; however, with cost-effective state-of-the-art technologies, the available number is restricted to a relatively small value (say, less than 30) for trunk transmission applications where transmission distance is large and high-quality transmission is required. The number of unidirectional fibers per link is also a parameter. Of course, in a large network accommodating relatively heavy traffic (path demands), multiple fibers are required per link when the number of wavelengths that can be multiplexed in a fiber is limited. In a LAN application, on the other hand, traffic demands (total traffic) are relatively small and unidirectional/bidirectional bus or ring topology using a single fiber for each direction can be utilized. Furthermore, network restoration with optical paths may be implemented, as mentioned in Section 4.4.3. Considering *paths* instead of *circuits*, a network will be designed so that all path demands are satisfied (paths are established/released based on long-term provisioning), while a certain blocking probability may be set for circuit establishment (call-by-call processing). Since optical paths are semipermanently established, the objective function is to minimize necessary network resources (necessary number of wavelengths or fibers in the network, or necessary cross-connect system scale), given that all path demands are to be completely satisfied. If there are not enough network resources, they will be added to accommodate the demands. On the other hand, a circuit network is created with call-by-call wavelength channels (wavelength channels are utilized for creating a circuit layer network), and some calls may be blocked when the available wavelength number is limited. In this application, the objective function will be to minimize call-blocking probability. In the following, we discuss WP and VWP accommodation design issues. Call-by-call WDM optical channel establishment issues are discussed in [115,120–123].

Single-Fiber Links and Unlimited-Wavelength Number

Some study results [115–117] are discussed for a network where each link consists of one unidirectional fiber (or two opposite direction unidirectional fibers when bidirectional path demands are considered) and the number of wavelengths per fiber is unlimited. For the network, the object function of path accommodation design is to maximize the wavelength utilization or to minimize the number of wavelengths required to accommodate a given set of WPs, assuming the physical network topology is given. The physical network topology is determined by a set of nodes and a set of unidirectional fibers between nodes.

In the VWP scheme, wavelengths are assigned link by link, so the number of wavelengths required for identifying all paths accommodated within each link equals the number of paths accommodated within the corresponding link. As a result, the number of wavelengths required in a VWP network equals the maximum number of paths accommodated within any one link. The number of wavelengths required for WP networks is generally more than that required for VWP networks. That is, the VWP scheme maximizes the degree of wavelength reuse in the network, so fewer wavelengths are needed than with the WP scheme. The extent of the wavelength reduction achieved with the VWP scheme depends on various conditions that include optical path restoration scheme (which will be explained in Section 4.5.2), physical network topology, traffic demands, and so on. The WP accommodation problem has been proven to be NP-complete, so there is no deterministic algorithm that can provide a solution in polynomial time [115]. Therefore, heuristic WP accommodation design algorithms that can solve the wavelength assignment problem must be established to enable large-scale WP network design. It has been found that by optimizing the route and wavelength assignment for WPs, the number of wavelengths required by the WP and VWP schemes are, in many cases, roughly equal when network restoration is not considered [115–119]. Regarding this, for networks with acyclic topologies (topologies that do not contain cycles; a tree network is an example), it has been proved that the lower bound on the number of wavelengths achievable by the WDM scheme is given by that in the VWP accommodation case [115].

Some WP accommodation heuristics applicable to a general topology network have been developed [115–118] that can be used to establish the number of WPs necessary where the objective function is to establish all the WPs demanded using as few wavelengths as possible. Figure 4.19 depicts a flowchart of VWP and WP accommodation algorithms [117]. The point of the optical path accommodation design algorithms is as follows. VWP accommodation design requires only path route optimization. In WP accommodation design, both path route and wavelength assignment must be optimized. In the WP accommodation algorithm shown in Figure 4.19(b), they are optimized independently. The

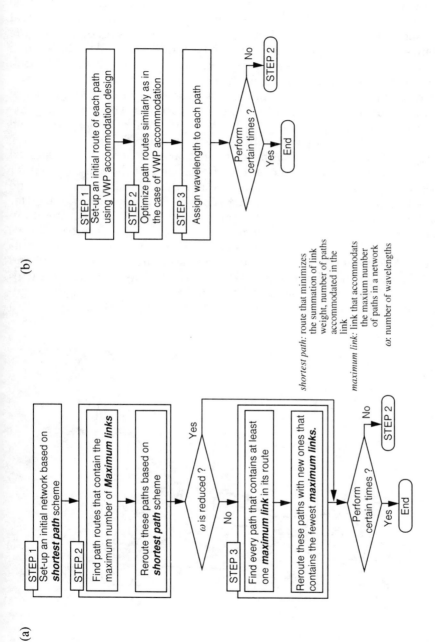

Figure 4.19 (a) VWP and (b) WP accommodation design algorithms. (*Source:* [117]. Reprinted with permission.)

details of the ideas and procedures of each process are presented in [117]. Some of the obtained results are described below. Figure 4.20 shows the physical topology used in the evaluation. It is a slightly modified version of the existing long-distance network in Japan and consists of 50 nodes. The total traffic volume (total number of paths) was set to values of 1,225, 1,838, and 2,450. For each traffic volume, three different random traffic patterns were applied (nine accommodation design models were tested). Table 4.7 shows the required number of wavelengths obtained by the heuristic methods. Regardless of the traffic volume, the number of wavelengths required by the WP and VWP schemes are shown to be almost equal when path restoration is not considered.

As depicted in Table 4.7, it has been shown that in most cases, the required number of wavelengths for WP accommodation is almost the same as that for VWP [115–119]. This can be explained by considering the lower bound on acyclic topologies. A discrepancy between WP and VWP can arise when cycles are contained in the physical network topology and some WPs exist that form cycles. This effect, however, can become small when a large number of WPs have to be established in the network. This is because the number of wavelengths required for VWP is determined by the most congested link in the network and the number gives the lower bound for a WP network (link utilization efficiencies of many of the other links can be small and the links allow wavelength assignment for WPs that should be accommodated within the links with a lot of freedom).

Figure 4.20 Physical network topology (*After:* [117]).

Table 4.7
Number of Wavelengths Needed for WP and VWP Schemes

Traffic Volume	Number of Wavelengths VWP	WP	$\frac{WP\text{-}VWP}{VWP}(\%)$
	206	210	1.90
1,225	221	221	0.00
	207	207	0.00
	304	315	3.62
1,838	310	311	0.32
	345	352	2.03
	403	416	3.22
2,450	404	415	2.72
	431	436	1.16

Note: Three traffic patterns were examined for each traffic volume.
After: [117].

Single-Fiber Links and Limited-Wavelength Number

When we consider *circuits* rather than *paths*, call-by-call wavelength channel accommodation will be necessary. The constraint of a single-fiber link and a limited-wavelength number is utilized in this application (not path accommodation). In this application, the call-blocking probability is an important network performance parameter. Several call-by-call WDM optical channel admission methods have been proposed [115,120–123]. In all of them the objective function is to minimize the call-blocking probabilities given that the available number of wavelengths in a fiber is limited and only single-fiber links are used.

The obtained results exhibit complex trends and depend on the network conditions, such as traffic load. In some cases, the average blocking probability with WPs (as mentioned above; although these are not paths but circuits, the words WPs and VWPs are used here for simplicity) is lower than with VWPs [115], although the WP scheme exhibits higher blocking probability than the VWP scheme when considering only long circuits. This unexpected result occurs for the following reasons. A long circuit takes up network resources that can be utilized by multiple short ones. The larger blocking probability for long circuits results in more unused network resources, which can be utilized by many short circuits and result in a lower blocking probability for short circuits. The number of the short circuits in the network is larger than that of the long circuit, and therefore the average blocking probability becomes smaller with WPs than with VWPs.

Another routing algorithm for wavelength channels corresponding to the WP and VWP schemes has been proposed. By implementing a limited number of wavelength conversion at call-by-call switch nodes, the call-blocking probability is improved [120]. It is shown that under light loading, the VWP scheme can reduce the blocking probability; however, at heavy loading the blocking probability of the VWP scheme can be larger than that of the WP scheme [123].

4.5.2 Path Accommodation Design Considering Network Restoration

Network Restoration With Optical Paths

Network failure restoration can be realized effectively with optical paths. Failure restoration is performed by cross-connecting failed paths to their backup paths. Each backup path can be pre-established in advance (network resources for restoration are reserved) of a failure; a specific network reliability can be guaranteed and the restoration time can be minimized. In the WP scheme, wavelengths must be preassigned along with routes to the backup paths so as to avoid wavelength collision. Another failure restoration strategy based on the real-time restoration path searching upon failure requires longer restoration time and so is not adopted here, although the scheme is effective for restoration of any failures, including multiple ones.

There are two restoration schemes in terms of restoration endpoints. One is restoration between failed link-fiber-terminating nodes. The other is between failed path-terminating nodes. The latter requires fewer spare resources for restoration at the cost of much more restoration processing since there can be many restoration node pairs for a single line/link failure and the restoration routes can be more distributed in the network. For long-haul trunk transmission network application, transmission cost is significant, so it is important to minimize spare capacity. In the following, therefore, only path-termination node restoration is considered.

Different restoration methods for WPs and VWPs are explained in Figure 4.21(a–c). Method 1 (Figure 4.21(a)) is for VWPs. When a failure is detected, failed paths are switched to the restoration paths, for which unused wavelengths are assigned link by link along each restoration path. For WP restoration, two restoration schemes have been identified: method 2 (Figure 4.21(b)) and method 3 (Figure 4.21(c)). Method 2 assigns to a backup path a wavelength that can be different from that assigned to the corresponding active path. In method 3, on the other hand, the same wavelength is assigned to a backup path as the corresponding active path. It is apparent that method 2 is

Photonic Transport Network 161

(a)
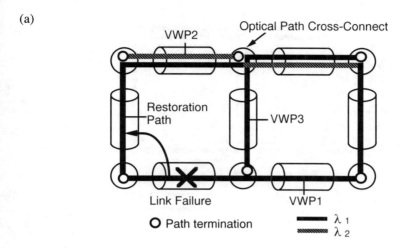

(b)
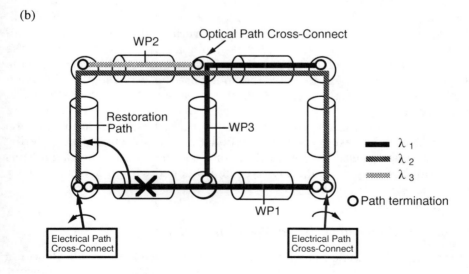

Figure 4.21 Restoration methods for (a) VWP (method 1), (b) WP (method 2), and (c) WP (method 3).

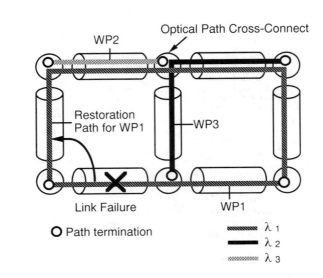

Figure 4.21 (continued)

more flexible than method 3 in terms of wavelength assignment for restoration paths and, accordingly, less spare resources for restoration paths are expected. However, in method 2, coordination with electrical path cross-connects is necessary, as shown in Figure 4.21(b), for assigning a different wavelength to the restoration path, although the restoration paths are optical. This is because the WP cross-connect system usually does not host the wavelength conversion function, as will be explained in Section 4.6. Note that in methods 2 and 3, the reserved resources (or reserved wavelengths for restoration purpose) in every link are commonly used for restoration against different failures in the network. As in the case of path accommodation design without considering network restoration (see Section 4.5.1), path accommodation design algorithms considering network restoration have been developed with different boundary conditions, as explained below.

Single-Fiber Links and Unlimited-Wavelength Number

For the case of a single-fiber link and unlimited wavelengths, an algorithm based on iterative improvement was developed [124,125]. Some of the results are demonstrated here. The physical network topology tested was a polygrid with $3 \times n$ nodes; this network topology allows a more general conclusion to be derived than is possible with a specific physical topology. For any one link failure in the network, 100% restoration was guaranteed. Each restoration path was established so that it did not share any links with the corresponding active path, since this greatly reduces the necessary database on the restoration paths and so enables the algorithm to be applied more easily to large-scale networks [124]. The path demands (the number of active paths between node pairs) were randomly set. Figure 4.22(a) shows the necessary number of wavelengths when restoration was considered; path demands were changed from 1.0 to 2.0 for the network with $n = 5$. The unit of path demand is defined as the total number of active paths established between all node pairs in a full-mesh configuration (i.e., $3 \cdot {}_nC_2 (= 105)$). As shown in Figure 4.22(b) method 2 and method 3 require approximately 10–13% and 26–29% more wavelengths, respectively, than method 1.

Figure 4.23(a) shows the required number of wavelengths for a polygrid network with $3 \times n$ nodes; n was changed. The path demand was set to be 1 ($3 \cdot {}_nC_2$ paths) and one active path was established between all node pairs in a mesh configuration for each network size. The ratio of the number of wavelengths required for WP (methods 2 and 3) to VWP (method 1) decreases as network size increases, as shown in Figure 4.23(b). This is because the

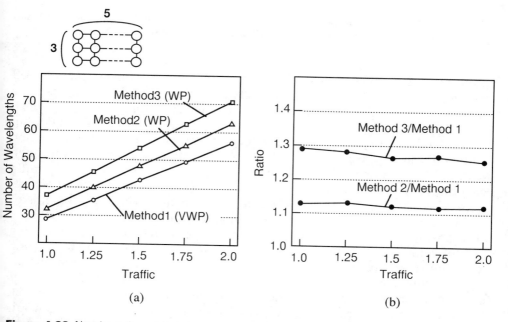

Figure 4.22 Number of wavelengths vs. traffic: (a) number of wavelengths and (b) ratio of number of wavelengths of WP to VWP (*After:* [124]).

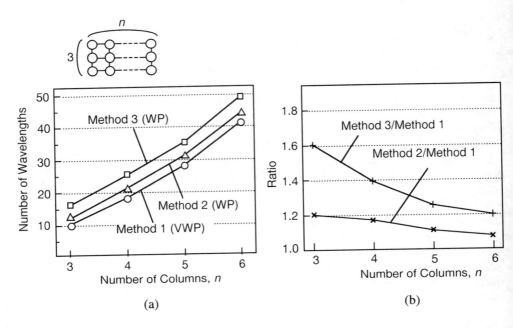

Figure 4.23 Number of wavelengths vs. network size: (a) number of wavelengths and (b) ratio of number of wavelengths of WP to VWP (*After:* [124]).

probability that different restoration paths can use, in common, a reserved wavelength in a fiber for restoration increases as network size increases, thus enhancing wavelength reuse probability among restoration paths.

Figure 4.24 shows another evaluation result using a heuristic algorithm for a 15-node nonregular physical topology network, which models the existing middle-distance network in Japan [117]. The required number of wavelengths with and without network restoration was evaluated. Five randomly set traffic patterns, whose traffic volume was constant ($105 = {}_{15}C_2$ paths for one direction) were accommodated using the WP and VWP schemes. The lower bounds of the required number of wavelengths with and without network restoration is also shown, which can be obtained without determining each path route. Note that this bound may not be attained [117]. The lower bound is useful for judging the degree of routing optimality, although it does not include any information on individual path routes. When the obtained optimal value equals the lower bound, the obtained value is assured to be the true optimum, although the lower bound might be smaller than the true optimum in certain cases. In general, obtaining a "good" lower bound (a lower bound that is very tight) tends to be very difficult when the network size becomes large. One procedure for obtaining

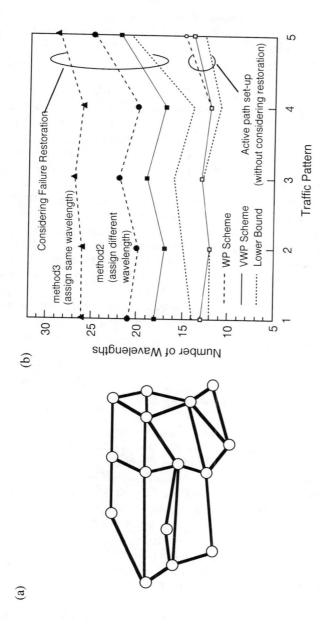

Figure 4.24 (a) Network model and (b) number of wavelengths against traffic pattern. (*Source:* [117]. Reprinted with permission.)

a lower bound is given in [117]. Other conditions used in the evaluation are basically the same as those explained above. As for the active path network (without considering failure restoration), the number of wavelengths required by the WP and VWP schemes almost equals the lower bound. Namely, there is only very small differences between the WP and VWP schemes, as explained before.

When failure restoration is considered, the differences between the required number of wavelengths and the corresponding lower bound are relatively large, even for the VWP scheme. This may be explained as follows. In the algorithms used [117], since restoration paths are selected so as to minimize the summation of link cost, circuitous routes that have excessive hopping are often not selected even though they may reduce the required number of wavelengths. However, this approach is practical because one important network requirement is to prevent the creation of the excessively long hopping paths that would make transmission and cross-connect node costs excessive.

It is also shown that the difference between the WP and the VWP schemes is enhanced when failure restoration is taken into consideration. Moreover, with the WP scheme, the difference between the two restoration schemes (methods 2 and 3) is also significant. This is partly due to the following reason. As explained before, the discrepancy between the WP and VWP schemes arises when cycles are contained in the physical network topology and WPs exist that form cycles. When path restoration is considered, active and the restoration paths are established at the same time. The end nodes for these two paths are the same, so circles are always created in method 3.

Multiple-Fiber Links (Limited Wavelengths)

Multiple fibers per link are inevitable when creating a large-scale network where traffic demands are high. This is because the number of wavelengths that can be multiplexed in a fiber is limited to a relatively small number [126,127], say, less than 30 with state-of-the-art technologies; the assumption is that the applications need high transmission quality even after long-haul transmission. WP and VWP accommodation design algorithms have therefore been developed for a network in which each link consists of multiple fibers [128–130]. Some of the results are demonstrated here.

For the algorithm developed in [130], the objective function is to maximize wavelength utilization efficiency in the links; in other words, to minimize the total number of WP/VWP cross-concoct ports (which consist of intraoffice and interoffice ports) required at each node. A different objective function—minimizing the number of WP/VWP cross-connect interoffice ports—may be adopted, if required, and this is done easily by slightly modifying the algorithm.

The generic optical path cross-connect system is depicted in Figure 4.25. The total number of incoming or outgoing ports of a WP/VWP cross-connect, hereafter called cross-connect scale, is given as N, the summation of interoffice port number as N_1 and intraoffice port number as N_2. The intraoffice ports are connected to electrical cross-connect systems in the node, for example. This value, N, is an important parameter for realizing economical cross-connect systems. Specific optical path cross-connect system hardware is discussed in the next section (Section 4.6).

The principles of the WP accommodation design algorithm [130] are briefly explained here. In VWP accommodation design, only the route need be determined. The fiber requirements of each link can be obtained from the number of accommodating paths and the maximum number of wavelengths per fiber. The first step is to perform active path accommodation design. The restoration paths are then accommodated as follows. First, a certain one-link-failure is assumed and the alternative routes for all failed active paths are located. This allows the fiber requirements for the said one-link-failure to be determined. This process is iterated until every one-link-failure is considered. Both the active path and restoration path routes are searched, utilizing the shortest path scheme with an appropriate weighting function so the number of underused fibers (fibers accommodating much fewer paths than the maximum number of wavelengths per fiber) is minimized.

In WP accommodation design, not only path routes but also path wavelengths must be simultaneously optimized. The wavelength assignment problem is known to be NP-complete as mentioned before. The route and wavelength assignment optimizing processes were separately conducted so as to simplify the algorithm. That is, wavelengths were assigned to the path routes established by the above VWP accommodation design process. The wavelength assignment process is iteratively performed on active WPs first; path routes are divided into the minimum number of groups such that each group must contain as many WP routes as possible and routes in the same group do not share any links of the physical topology with each other. Then, fibers and wavelengths are assigned heuristically group by group. The restoration path wavelengths are selected so that additional fiber requirements are minimized. The next step is to calculate the average WP cross-connect system scale of all WP cross-connects in the network, and iterate the wavelength and fiber assignment process a certain number of times and adopt the group arrangement that gives the minimum average cross-connect system scale as the optimal one. Then, restoration WPs are set up [130].

Some of the accommodation design results are explained here. The physical network topology used for the evaluation is the same as shown in Figure 4.24(a). Figure 4.26(a) shows the total number of fibers needed by the WP and VWP schemes where the number of active paths to be established is

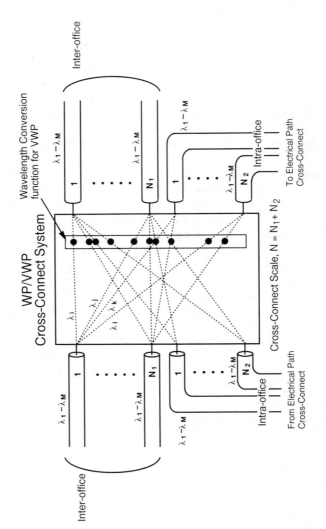

Figure 4.25 Generic optical path cross-connect architecture.

changed. The total number of fibers includes interoffice fibers and intraoffice fibers accommodating paths terminating at nodes. Pairs of bidirection paths (for upstream and downstream paths), the routes of which are the same, are considered. The assumed traffic distribution pattern between node pairs is random, and ten different patterns were tested. The total number of optical fibers shown Figure 4.26(a) is the average value for the ten different traffic patterns. Method 2 was adopted as the WP restoration method, as explained in Figure 4.21(b); a restoration path wavelength can be different from the corresponding active path wavelength. The restoration condition is the same as explained before: 100% restoration for any one-link failure in the network. The number of wavelengths that can be multiplexed into a fiber is restricted to four or eight. Figure 4.26(b) shows the ratio of total number of fibers for the WP and VWP schemes. It is shown that for 4 and 8 wavelengths per fiber, the WP scheme requires 16% and 24% more fibers, respectively, than the VWP scheme when path restoration is taken into consideration. Thus, the difference in the total required number of fibers between the WP and VWP schemes increases when more wavelengths can be multiplexed within each fiber.

Figure 4.27(a) shows the average number of cross-connect ports in the network, and Figure 4.27(b) shows their ratio for the WP to VWP schemes. It is shown that when the wavelength limitation is 8, the average number of cross-connect ports required per WP node is about 45 when the active path number established in the network is 840 and path restoration is considered. Note that in this cross-connect node scale evaluation, all paths are accommodated within cross-connect systems at every intermediate node along the paths; however, in some cases some of the paths may not be so accommodated. In other words, some paths can be direct and are not accommodated within cross-connect systems along the paths, or accommodated within a limited number of cross-connect nodes along the paths. The requirements depend on the path-establishment strategy, which will be determined considering cross-connection cost and transmission cost ratio and the required network reliability. Therefore, the cross-connect port number demonstrated in Figure 4.27 gives the maximum value for the network at a given active path demand.

For the model network shown in Figure 4.28, which consists of 40 nodes, Figure 4.29 shows other evaluation results on the total number of required fibers. In the evaluation, the number of wavelengths that can be multiplexed in each fiber was changed from 4 to 32. The total number of active paths to be established in the network is set to be 1,560 (= $2 \times {}_{40}C_2$), which corresponds to the number of paths required to create a full-mesh connection (for upstream and downstream paths) among nodes. The assumed traffic distribution pattern between node pairs was randomly set and ten different patterns were tested. The total number of required optical fibers in the network shown in Figure 4.29(a) is the average value for the ten different traffic patterns. The number

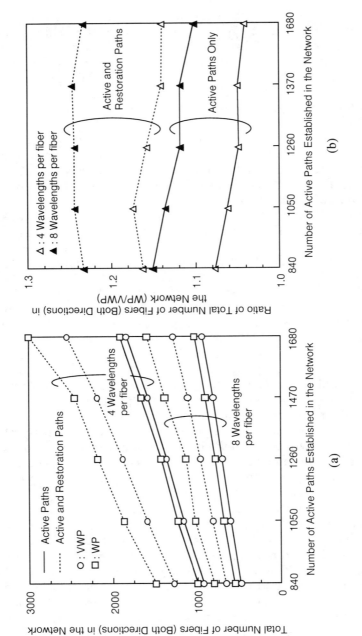

Figure 4.26 (a) Total number of fibers in the network and (b) ratio of total number of fibers in the network.

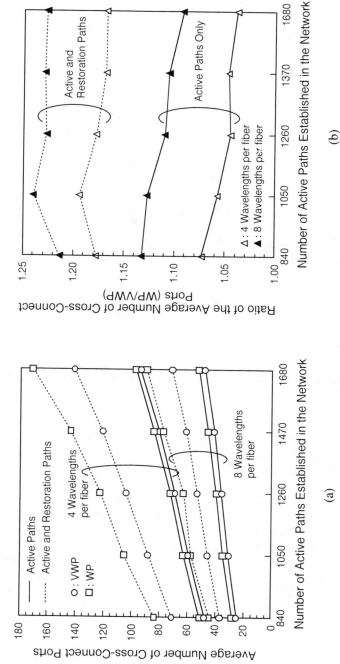

Figure 4.27 Optical path cross-connect scale: (a) average number of cross-connect ports and (b) ratio of average number of cross-connect ports.

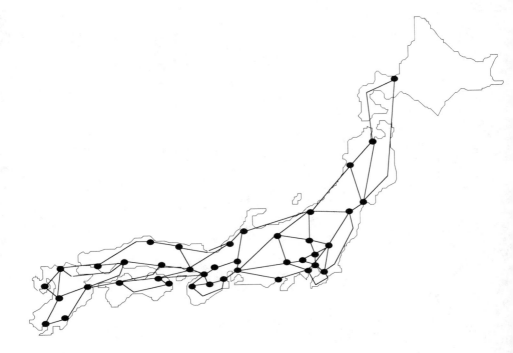

Figure 4.28 Physical network topology (*After:* [128]).

of fibers does not decrease linearly against the number of wavelengths. This is because the average utilization efficiency of a fiber decreases as the number of wavelengths that can be multiplexed in each fiber increases when path demands are constant. Therefore, increasing the maximum number of wavelengths multiplexed in a fiber is not always effective in terms of fiber utilization efficiency. Figure 4.29(b) shows the ratio of the total number of fibers with the WP and VWP schemes. As shown in the figure, the relative inefficiency in WP networks increases with the upper bound of the wavelength number. When the wavelength limitation is 32, the ratio of total number of fibers is larger for active paths than active plus restoration paths. This is because, considering restoration, the total number of paths to be established in the network increases, and the fiber utilization inefficiency imposed by the larger number of wavelengths that can be multiplexed in each fiber is mitigated. Figure 4.30(a) shows the average number of cross-connect ports in the network, and Figure 4.30(b) shows their ratio for the WP and VWP schemes. The same tendencies are obtained as for the number of fibers.

As discussed herein, the total number of fibers and cross-connect ports required in the network with the WP scheme can be larger by 20% or more

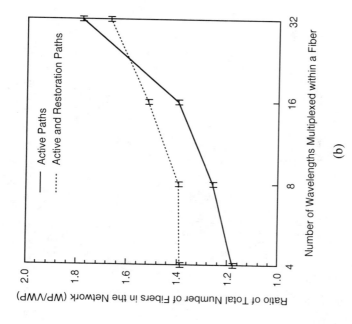
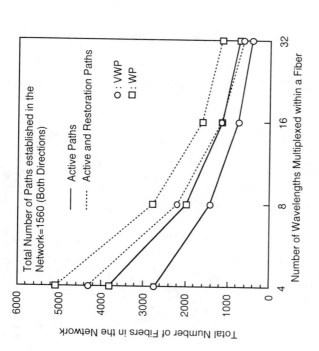

Figure 4.29 (a) Total number of fibers in the network and (b) ratio of total number of fibers in the network.

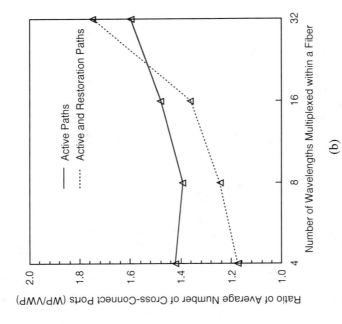
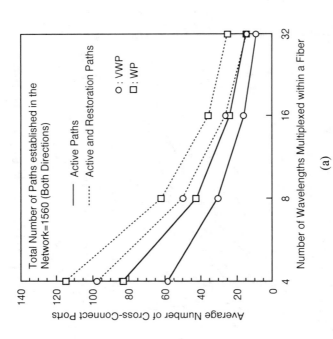

Figure 4.30 Generic optical path cross-connect scale: (a) average number of cross-connect ports and (b) ratio of average number of cross-connect ports.

than with the VWP scheme when network restoration with optical paths is taken into account. The transport cost consists of transmission cost, cross-connect node cost, and operation cost. In application to the backbone network or to a large physical network, transmission cost can be a major cost, so the 20% transmission cost increase can be considerable. However, in a more practical assessment, the cost difference can be mitigated since the average link utilization in the network cannot always approach 100% (in the existing trunk transmission networks, the average link utilization would be 80% or less). This is partly due to the granularity of newly added fiber cables to existing routes or new routes to satisfy traffic demand increases. Each addition is achieved by laying one or more fiber cables, each of which may consists of 100 or more fibers. The number of fibers in a cable will tend to be fairly large considering recent increases in construction costs. This large fiber step size will lower the average link utilization in the network and effectively reduce the impact of the possible link utilization inefficiency associated with the WP scheme.

The increased generic link utilization inefficiency and the increased cross-connect port demand that can occur with the WP scheme, and the significant WP advantage of much simpler cross-connect architecture, are important criteria in determining the applicability of the WP scheme. The significance of the disadvantage incurred with the WP scheme strongly depends on the application area and has to be evaluated considering various application-specific conditions and technologies available at the time.

In the next section, we will discuss another important issue in realizing the optical path network—architecture and technology of WP and VWP cross-connects, where the inherent WP cross-connect advantage will be evaluated.

4.6 OPTICAL PATH CROSS-CONNECT

4.6.1 Generic Optical Path Cross-Connect System Architecture

The generic optical path cross-connect node architectures for the WP and VWP schemes are shown in Figure 4.31 and 4.32. For the optical path cross-connect systems depicted in Figure 4.31, some of the input and output ports are used for intraoffice interfaces with electrical facilities such as electrical path cross-connect systems. Those shown in Figure 4.32 utilize optical add/drop multiplexers, which provide intraoffice interfaces. The architecture shown in Figure 4.32 is preferred when an add/drop multiplexer and a cross-connect system that provides only interoffice interfaces can be developed with less hardware than the cross-connect system that provides both inter and intraoffice interfaces. The functional difference between WP and VWP cross-connect systems is the wavelength conversion function required in the VWP cross-connects. In these figures, wavelength conversion is placed just before the optical space-

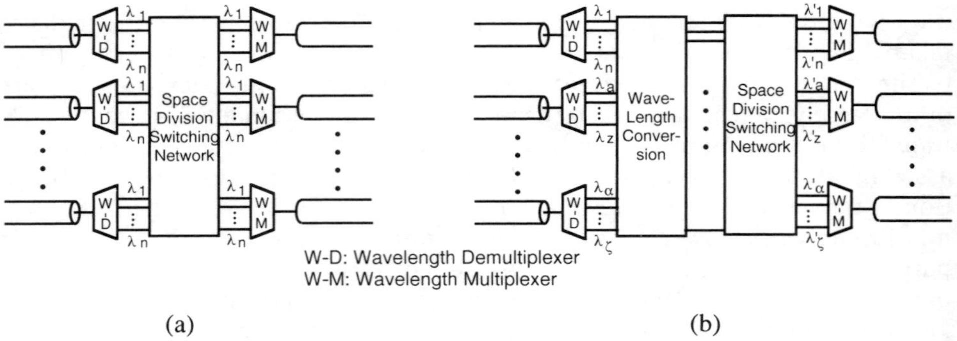

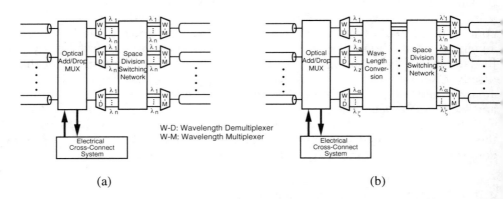

Figure 4.31 Generic optical path cross-connect node architectures: (a) WP and (b) VWP (*After:* [72]).

Figure 4.32 Generic optical path cross-connect node architectures with optical add/drop multiplexers: (a) WP and (b) VWP (*After:* [145]).

switching network. This configuration is called here the prefix type. With this architecture, tunable-wavelength conversion is necessary. On the other hand, it is possible to place wavelength conversion just after the optical space-switching network, the postfix-type configuration. This type converts the wavelength to a fixed wavelength specific to each converter.

The wavelength conversion function may be performed directly in the optical domain in the future with the differential frequency generation technique [131], four-wave mixing techniques in laser diodes [132,133] and in fibers [134–136] or bistable laser diodes [137]; however, these technologies are still immature. Wavelength conversion via electrical regeneration is very feasible considering available techniques. Integrated optical-to-electrical (O/E) and electrical-to-optical (E/O) circuits offering this function have been developed [138]. Wavelength conversion permits the optical power level of each VWP to be individually adjusted before the VWPs are multiplexed into a fiber. This greatly facilitates the design of the optical power level and minimizes the crosstalk problem [139–142]. The crosstalk problem among VWPs becomes much more significant [139] if the optical power levels are very different. Such differences can be created, if power level adjustment is not employed, by the wavelength-selective loss of the optical components used in the network and the different numbers of links and nodes traversed by each VWP.

Concerning WP cross-connects, two approaches to optical repeaters are possible: one is to employ optical regeneration at the electrical level, as explained above, and the other utilizes nonregenerative repeaters (i.e., optical amplifiers). Optical regeneration at the electrical level is feasible and practical not only for the VWP scheme, but also for the WP scheme, especially for global area network applications where optical transmission loss is severe due to long-haul transmission and high transmission qualities are required [71,72]. By utilizing optical amplifiers, on the other hand, we can attain the maximum transparency of the transmission signal. In this architecture, however, crosstalk generated within optical devices and amplified spontaneous emission (ASE) noise of optical amplifiers accumulates, which limits the application distance [143].

The introduction of optical regeneration at the electrical level restricts the transmission signal attributes such as bit rate and signal format, analog or digital. However, the effects of the restriction can be mitigated by using a cross-connect system architecture designed to support different transmission signals by replacing only the O/E and E/O cards, leaving other major hardware portions unchanged, as will be explained in Section 4.6.2.

When optical regeneration at the electrical level is adopted, the WP cross-connect architecture and the postfix-type VWP architecture become very similar, although the required number of components (space switches, or star couplers and wavelength multi/demultiplexers) differ. Such a cross-connect system can be realized with fixed-wavelength LDs, instead of wavelength tunable LDs, as depicted in the next section.

4.6.2 Optical Path Cross-Connect Switch Examples [144]

Switch Fabric Using SD Switches

Space division (SD) switches can be used for the optical space-switching network shown in Figure 4.31. The following discussion assumes N incoming and outgoing ports and M wavelengths multiplexed into each fiber. Several SD-switch-based architecture have been proposed [145,72]. Figure 4.33 shows one

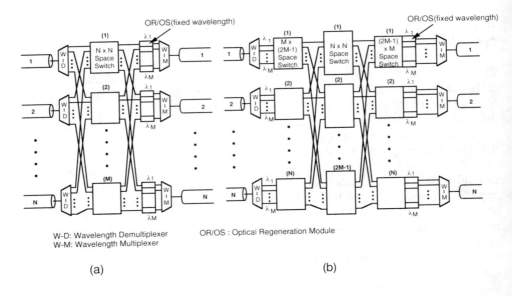

W-D: Wavelength Demultiplexer   OR/OS : Optical Regeneration Module
W-M: Wavelength Multiplexer

(a)   (b)

Figure 4.33 WP/VWP cross-connect switch architecture based on space division switches: (a) WP and (b) VWP.

example where optical regeneration at the electrical level is utilized, as is the postfix type architecture. For WP cross-connects, the regeneration part can be replaced with optical amplifiers positioned after incoming fibers and/or before outgoing fibers, if necessary. In the WP scheme, optical cross-connection is performed while keeping the wavelength constant; therefore, an $N \times N$ SD switch is utilized for each optical path (wavelength channel) to select the output port, as shown in Figure 4.33(a). On the other hand, in VWP cross-connects, the required SD switch size is $NM \times NM$, and it should be nonblocking in a strict sense, as will be discussed in Section 4.6.3. For constructing a large $NM \times NM$ switch, a nonblocking multistage switching network architecture, like the Clos switching network [146], is employed as illustrated in Figure 4.33(b). These cross-connect systems are postfix type, so fixed-wavelength lasers are used as the optical senders (OSs). The major benefit of this SD-based VWP cross-connect architecture is that it consists of fixed-wavelength LDs instead of tunable LDs. However, if tunable-wavelength LDs or some equivalents are available, the multistage switching network for the VWP can be simplified [147], as shown in Figure 4.34.

The ratios of VWP hardware amount to WP and VWP/WP needed to construct a large SD switch for postfix-type WP and VWP optical cross-connect systems [145] are shown in Figure 4.35. Figure 4.3.5 assumes that the maximum

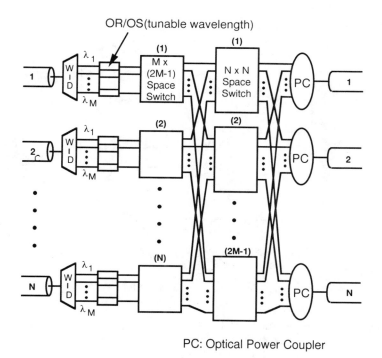

Figure 4.34 Prefix type VWP cross-connect switch architecture based on space division switches (*After:* [147]).

component switch size is 8 by 8 and that the hardware amount of each component SD switch is of the order of n^2 where n is the switch size of each component SD switch (the maximum is 8). Order n^2 corresponds to the state-of-the-art optical switch technology (for example, the cross-bar matrix switch). There are many ways of constructing a multistage Clos switching network with small SD switches. The evaluation used the best combination. It is shown that the ratio VWP/WP ranges from 2.9 to 5.5 when N is larger than 16. This is a significant difference when N is large, where total SD switch hardware becomes extremely large. Therefore, this postfix-type VWP architecture is not practical with state-of-the-art technologies when N is large, although it can be realized with fixed-wavelength LDs.

Another WP cross-connect architecture [148,149] is shown in Figure 4.36. In the architecture, interoffice and intraoffice interfaces are clearly distinguished, and therefore the maximum number of WPs that can be terminated (originated) at the node is predetermined (for Figure 4.36, it is four). When

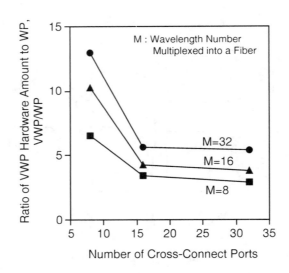

Figure 4.35 Space division switch hardware requirement for VWP/WP.

tunable-wavelength filters are used instead of fixed-wavelength filters in front of SD switches as in Figure 4.36, the same wavelength WPs on different incoming fibers can be terminated at a node. With this configuration, however, this architecture does not offer strictly nonblocking characteristics, but rearrangeable nonblocking ones. The configuration may then limit the application area, as will be addressed in the next section.

Switch Fabric Using Optical Power Splitters and Filters

The space switches explained thus far can be replaced with a combination of optical power splitters and (tunable) wavelength filters [145]. In this architecture, requirement differences between WP and VWP cross-connect switches become smaller than with SD-based switches. Figure 4.37 shows such WP and VWP architectures; multiwavelength selective filters (MWSFs) that can select any combination of wavelengths and tunable wavelength filters (TWFs) that can select any one wavelength are combined with optical power splitters [145]. Different types of MWSFs—such as those that use acousto-optic interaction [150], or a combination of a wavelength demultiplexer, optical switches, and a wavelength multiplexer—have been reported. As shown in Figure 4.37, each input optical path is routed to its appropriate destination port by controlling the MWSF. In the VWP cross-connect, the first-stage MWSF, followed by the TWF, leads the input optical path to the *appropriate* wavelength

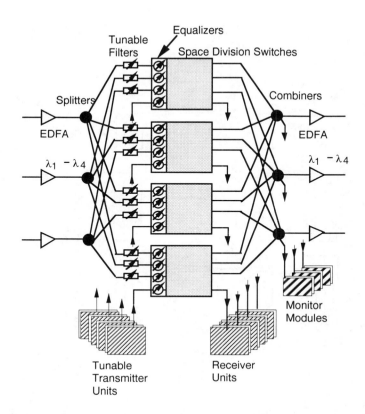

Figure 4.36 SD-based cross-connect switch architecture example (*After:* [148]).

onverter, and the last-stage MWSF routes the converted paths to the destination ort. *Appropriate* here means that paths of the same wavelength are never ed to the same star coupler. The VWP cross-connect switch shown here is a earrangeable nonblocking switch. The strictly nonblocking switch architecture hown in Figure 4.38 is composed of optical star couplers, gate switches ($N \times 1$ SD switches) and TWFs [151]. This is a postfix-type optical cross-connect witch. For these switching networks, the key components are the MWSF and 'WF. The development of high-quality but very cost-effective MWSFs/TWFs s required to realize practical large-scale optical path cross-connect systems, ince at least $N \times M$ of them are needed.

witch Fabric Using DC Switches

igure 4.39 shows the schematic configuration of the WP and VWP crossonnect switch architectures that are based on delivery and coupling switches

182 Advances in Transport Network Technologies: Photonic Networks, ATM, and SDH

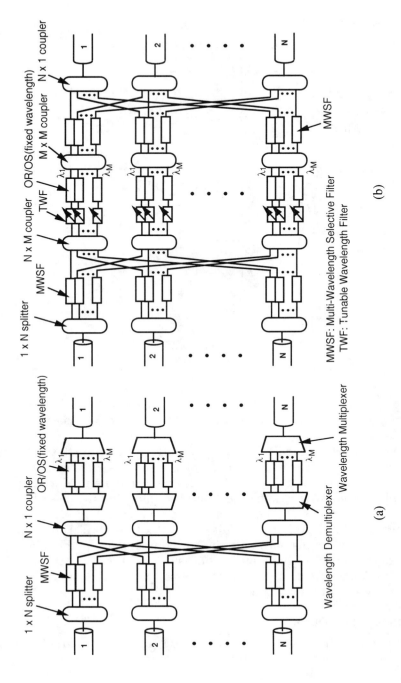

Figure 4.37 Cross-connect switch architecture using MWSFs: (a) WP and (b) VWP (*After:* [144]).

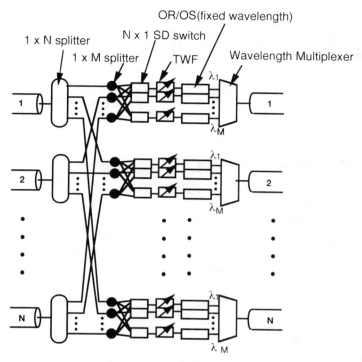

TWF: Tunable Wavelength Filter

Figure 4.38 VWP cross-connect switch architecture using a parallel λ-switch (*After:* [151]).

(DC switches) [152–154,147]. This switch is strictly nonblocking. The WP cross-connect switch with optical regeneration uses fixed-wavelength LDs while tunable-wavelength LDs are used for VWPs. Optical regeneration at the electrical level, optical receivers/optical senders (ORs/OSs) in Figure 4.39(a), can be eliminated from the WP cross-connects, which do not need optical regeneration (optical amplifiers may be added to the appropriate position when necessary).

Figure 4.39(b) shows the generic configuration of an $M \times N$ DC switch where optical couplers are utilized in common for VWPs and WPs, or W-MUXs can be utilized for WPs in order to minimize the coupling loss. The switch allows any of the M incoming optical signals to be connected to any of the N outgoing ports. This is not possible with the conventional $M \times N$ SD switch. Of course, optical paths with the same wavelength must be connected to different outgoing ports. Thus, if optical couplers are utilized in the $M \times N$ delivery and coupling switch, the cross-connect system architecture allows easy evolution

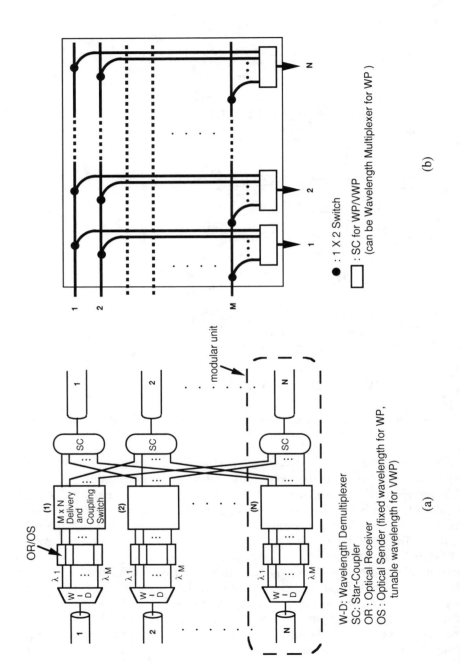

Figure 4.39 (a) WP/VWP cross-connect switch architecture using DC-switches and (b) generic configuration of $M \times N$ DC-switch

from the WP cross-connect to the VWP cross-connect by simply replacing the fixed-wavelength LDs in the OS modules with tunable-wavelength LDs, while leaving other major portions unchanged. In other words, this switch architecture provides the maximum commonality between WP and VWP cross-connects. This switch architecture also provides excellent modular growth capability, as will be further discussed in the next section.

4.6.3 Optical Path Cross-Connect Requirements

The requirements placed on optical path cross-connect systems for use in large scale optical path networks are as follows:

1. Strictly nonblocking for utilization in network restoration;
2. Maximum modular growth capability;
3. Easy evolution from WPs to VWPs;
4. Low crosstalk and low optical loss;
5. Based on rather mature and credible optical technologies.

Each item is explained and comparisons of the switch architectures described in the former section are discussed below.

Nonblocking Characteristics

A switch is nonblocking if any input can be connected to any unused output. Here, blocking does not include the blocking caused by wavelength collision between input and output ports, which can occur in a WP scheme. The nonblocking characteristics are classified [155] into strictly nonblocking, nonblocking in a wide sense, and rearrangeable nonblocking. A switch is strictly nonblocking if any input can always be connected in any viable way to any unused output without disturbing existing connections. A switch is nonblocking in a wide sense if an algorithm exists for setting up the connection in a way that guarantees that any future connection can always be established without requiring rearrangement of existing connections. A switch is rearrangeable nonblocking if all permutations are possible, but some existing connections may need to be broken and rearranged to allow the new connection to be established. The characteristics of strictly nonblocking or nonblocking in a wide sense are necessary for the flexible reconfiguration of path connections without affecting any active path already established. Thus, these characteristics are necessary when optical path cross-connect systems are used for network restoration.

In the cross-connect architecture explained in the previous section, the SD-switch-based WP cross-connect architecture shown in Figure 4.36, when it is used so that the same wavelength WPs on different incoming fibers can be terminated, and the VWP cross-connect architecture using MWSFs shown in Figure 4.37(b) are rearrangeable nonblocking, and the other cross-connect switches are strictly nonblocking, as summarized in Table 4.8.

The requirement of strict nonblocking entails large-scale SD switches in the SD-switch-based architecture, especially for a VWP cross-connect. If incoming and outgoing fiber numbers are N, and the number of wavelengths multiplexed into each fiber is M, then an $(N \times M) \times (N \times M)$ SD switch is required for the VWP cross-connect system (see Figure 4.32). To construct such a large-scale SD switch, a nonblocking multistage Clos switching network [146] should be used, as explained before. For a postfix-type VWP cross-connect, shown in Figure 4.33(b), if N is 16 and M is 8, a 128 by 128 SD switch is required. The switch can be constructed as a three-stage, 8 by 15(16)-16 by 16(15)-15 by 8(16), switching network (the number in parentheses denotes the required number of switches). If an 8 by 8 unit switch can be produced (this is possible with available technologies such as the planar lightwave circuit (PLC) technology, which will be addressed later in this section [156]), the 8 by 15, 16 by 16, 15 by 8 component switches should be further decomposed to 1 by 2(8)-8 by 8(2), 2 by 3(8)-8 by 8(3)-3 by 2(8), and 8 by 8(2)-1 by 2(8), respectively. In this case, a total of 109 (= 32 + 45 + 32) 8 by 8 switches are required. As evaluated here, when N is large, the SD switch hardware required for the postfix-type VWP cross-connect tends to be excessive for practical realization.

Modular Growth Capability

Another important design parameter is the modular growth capability of the cross-connect switch, as discussed in Section 4.4.3. Maximizing this parameter permits minimum initial investment, while allowing investment to be incrementally increased as traffic demand increases. It permits the economical introduction of optical path cross-connect systems to nodes with scales ranging from small to large. In terms of modular growth capability, two different levels of modularity can be identified: the number of wavelengths (number of WPs/VWPs) multiplexed in a fiber (wavelength modularity) and the number of input and output fiber pairs (link modularity). When the available number of wavelengths multiplexed into a fiber is relatively large, the wavelength modularity may also be important; however, the link modularity is more important when the number of multiplexed wavelengths is limited to the order of ten. Thus, only link modularity is discussed here.

Among architectures discussed so far, the SD-switch-based architecture shown in Figure 4.33 has no link modularity in terms of the SD-switch portion.

Table 4.8
Comparison of Optical Path Cross-Connect Architectures

WP/VWP	Architecture		Figure	Nonblocking Nature	Link Modularity	WP to VWP Commonality	Passive Component Loss
WP	SD-switch-based	Prefix-type		Strictly nonblocking	Low	No	Small
	SD-switch-based	Postfix-type	Figure 4.33(a)	Strictly nonblocking	Low	No	Small
	SD-switch-based	Tunable-wavelength filter	Figure 4.36	Strictly/rearrangeable nonblocking*	Low	No	Small
	Using MWSFs	MWSF	Figure 4.37(a)	Strictly nonblocking	High	No	Large
	DC-switch-based		Figure 4.39	Strictly nonblocking	High	Good	Medium-small
VWP	SD-switch-based	Postfix-type Fixed-wavelength OS	Figure 4.33(b)	Strictly nonblocking	Medium	Poor	Large
	SD-switch-based	Prefix-type Tunable-wavelength OS	Figure 4.34	Strictly nonblocking	Medium	Poor	Large
	Using MWSFs	MWSF Fixed-wavelength OS	Figure 4.37(b)	Rearrangeable nonblocking	High	Poor	Very large
	Parallel λ-switch	Postfix-type Tunable-wavelength filter Fixed-wavelength OS	Figure 4.38	Strictly nonblocking	High	Good	Medium
	DC-switch-based	Prefix-type Tunable-wavelength OS	Figure 4.39	Strictly nonblocking	High	Good	Medium-small

*See text.

The SD switch architecture makes the total number of SD switches equal the anticipated maximum number of input/output fibers even when the initial number of input/output fibers is small. Thus, the major portion of the cross-connect switch passive hardware (SD switches) is needed at the initial introduction stages. In other words, the SD-based architecture can be cost-ineffective. Both the DC-switch-based architecture (Figure 4.39) and the parallel λ-switch (Figure 4.38) demonstrate the highest level of link modularity. The MWSF-based architecture lies between them. The link modularity of the DC-switch-based architecture is explained in Figure 4.39(a), where the unit of addition/deletion (building block) is delimited with a broken line. As mentioned, the importance of link modularity is emphasized, however, a DC-switch-based cross-connect switch architecture that offers wavelength modularity has been developed as shown in Figure 4.40 [153]. (In this configuration, $N \times N$ DC switches are utilized; they can be replaced with conventional $N \times N$ SD switches in the WP cross-connect system.) The modular growth capability is compared in Table 4.8

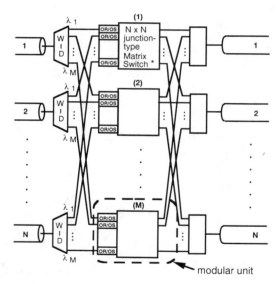

OR : Optical Receiver
OS : Optical Sender (fixed wavelength for WP, tunable wavelength for VWP)

☐ : SC for WP/VWP
(can be Wavelength Multiplexer for WP)

*: Matrix Switch consists of SC

Figure 4.40 WP/VWP cross-connect switch architecture using DC-switches with modularity in regard to the number of wavelengths (*After:* [153]).

An evaluation example [157] of the modular growth capability of two types of cross-connect hardware in terms of cost is shown here. Figure 4.41 (upper figure) compares relative hardware cost of the SD switch (Figure 4.33(a) for WPs and Figure 4.34 (for VWPs) and the DC switch (Figure 4.39)-based architectures in terms of the number of cross-connect ports. The cost includes all optical components needed for the optical cross-connect system such as optical space switch, optical regeneration, optical amplifiers, wavelength multi/demultiplexers, and optical power couplers. It does not include the switch control function cost, optical path OA&M cost, and intraoffice interface cost. It is shown that the DC-switch-based architecture exhibits better modular growth capability than the SD switch architecture. Since the port number needed by each optical path cross-connect in the network varies, this modular growth capability was evaluated in terms of total node cost in a network. Figure 4.41 (lower) shows an example of the electrical cross-connect port number

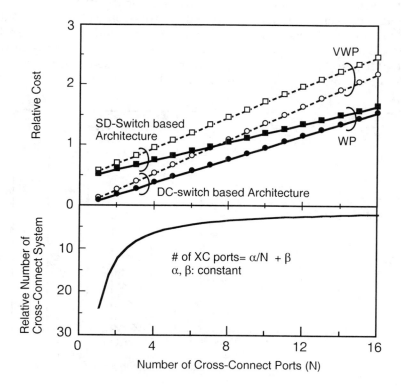

Figure 4.41 Cross-connect cost and relative number of cross-connects in the network. (*Source:* [157]. Reprinted with permission.)

distribution of Japan's long-haul transport network. The distribution is empirically expressed as

$$\text{cross-connect port number} \propto \alpha/N + \beta,$$

where α and β are constant. Figure 4.42 shows cost evaluation results for WP and VWP schemes, where α and β are assumed to be 1 and 0.03, respectively, considering an existing network. It is shown that in this example, 35–40% less cost is required with the DC-switch-based architecture than the SD-switch-based architecture for the network considered.

Evolution from WPs to VWPs

As discussed in Section 4.4.4, VWPs are more flexible than and have various advantages over WPs; however, the most significant VWP weakness is the wavelength conversion function required in cross-connect systems. With continued progress in device technologies, cost-effective and practical wavelength-tunable LDs or some equivalents are expected to become feasible in the not-so-distant future [158]. Especially for the DC-switch-based architecture, upgrading a WP cross-connect to a VWP cross-connect is simply performed by replacing

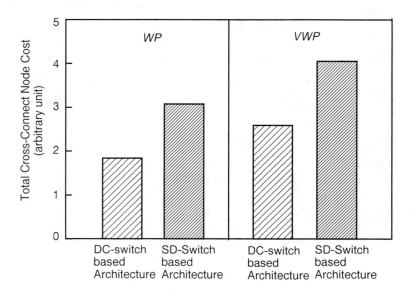

Figure 4.42 Total cross-connect node cost in the network. (*Source:* [157]. Reprinted with permission.)

the fixed-wavelength OSs with tunable ones. The parallel λ-switch-based architecture permits similar upgrading. It is apparent that the tunable-wavelength filters of the VWP cross-connect shown in Figure 4.38 can be replaced with fixed-wavelength filters to realize a WP cross-connect. On the other hand, other architectures discussed herein have no upgradability, so the commonality between WP and VWP cross-connects is very small (see Table 4.8).

Cross-Connect Switch Loss

The optical loss of the cross-connect system is an important parameter that determines the signal to noise ratio (SNR). The loss has to be compensated with optical amplifiers or through electrical regeneration. Optical-fiber amplifiers are suitable for amplifying wavelength-multiplexed signals(paths) simultaneously. Unfortunately, they are relatively large and can be costly to utilize many of them. The SD-switch-based WP cross-connect architecture exhibits lower loss than the others; the SD-switch-based VWP cross-connect exhibits relatively large loss. As shown in Figure 4.37(b), the VWP cross-connect architecture based on MWSF shows significantly large loss when its scale becomes large, since the total loss of optical couplers is at least $10 \times \log(N \times M \times M \times N)$ [dB]. When $M = 8$ and $N = 16$, for example, the loss is 42 dB. The actual loss of each architecture depends on the device technologies used. Here, some loss evaluation results are shown assuming planar lightwave circuit (PLC) technologies.

Figure 4.43 compares the total passive optical component loss against the number of incoming and outgoing fiber pairs (denoted as port number) between the SD-switch-based (Figure 4.33(a) for WPs and Figure 4.34 for VWPs) and the DC-switch-based (Figure 4.39) architectures for WP and VWP cross-connect systems [159,157]. The evaluations assumed $M = 8$, the use of PLC technologies, that the PLC waveguide loss is 0.15 dB/cm, and the connection loss between PLC circuit and multifiber (8 fiber) tapes is 1.0 dB (for both input and output fibers). All the losses of the optical connectors and splices between components and optical boards are included, as indicated by board and assembly designs. In the loss evaluation, the worst expected loss value of each optical component is assumed and the losses are simply added; no statistical treatment is applied.

The DC-switch-based architecture is based on forming one $M \times N$ DC switch (Figure 4.39(b)) on one Si wafer. The limitations of the present state-of-the-art technology, however, require the division of the switch into several components. The division of switch functions can be optimized considering the port number permitted by current technology. Regarding the SD-switch architecture, the Clos network has been optimized for the number of input/output ports by determining the optimum combination of each component switch in terms of the amount of hardware [145]. The scale of the 8x8 component

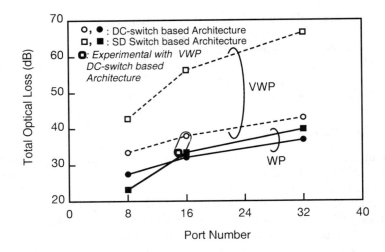

Figure 4.43 Comparison of total optical path cross-connect loss. (*Source:* [157]. Reprinted with permission.)

SD switch [156] is assumed to be maximum, since an 8x8 switch is already commercially available. As shown in Figure 4.43 for WP cross-connects, the DC-switch-based architecture exhibits almost the same optical loss as the SD-switch-based architecture; however, for VWP cross-connects, the DC-switch-based architecture attains significantly lower loss. The experimental value for the DC-switch-based VWP architecture is also included in Figure 4.43, when N is 16. The experimental details are given in [160]. This value is smaller than the worst expected loss value (38.1 dB) evaluated above since the expected value is just the simple summation of the worst loss of each component, as mentioned before. These optical losses, which stem from the passive optical components, must be compensated with fiber amplifiers.

Available Optical Technologies

To create a practical network system, rather mature and credible optical technologies are necessary. One example is to exploit already established transmission technologies utilizing direct intensity modulation of laser diodes. However, in the future, coherent modulation/detection technologies may be applied to more effectively realize equivalent tunable-wavelength filters [161,162], as required by the λ-switch architecture.

Recently developed PLC technologies are also important because they are very suitable for mass production and enable low-cost passive optical devices.

The PLC technologies use silicon substrates and silica glass waveguides. Their attractive features include [163–165]:

- They have large, low-cost substrate areas.
- The refractive index of the waveguides (silica glass) matches that of optical fiber, so their splicing losses are minimized.
- Computer-aided design (CAD) methods can be fully utilized to create a variety of functions.

However, PLC-based switches utilize thermo-optic effects to alter the refractive index of the waveguide, and relatively large switching power is required. Figure 4.44 shows the maximum switching power consumption of WP and VWP cross-connect switches, based on SD switches (Figure 4.33(a) for WP and Figure 4.34 for VWP) and DC switches (Figure 4.39). The evaluations assumed $M = 8$. Both SD and DC switches are composed of thermo-optic Mach-Zehnder interferometer 2x2 switch components [156]. Each 2x2 switch is activated by applying electrical power to the thermo-optic phase shifter located on the shorter waveguide arm of the Mach-Zehnder interferometer so that a π phase shift occurs. The typical power consumption is 0.45W for the on state [160]. As shown in Figure 4.44, the WP switch power consumption is comparable for both types of switches; however, for VWP switches, the DC-switch-based architecture needs much less power than the SD-switch-based

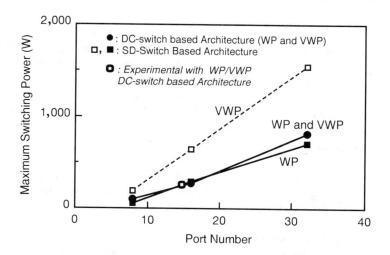

Figure 4.44 Maximum PLC switching power consumption. (*Source:* [157]. Reprinted with permission.)

architecture. The experimental value for a DC-switch-based WP/VWP cross-connect switch is also given in Figure 4.44, when N is 16. The experimental value was 257.3W, which is very close to the expected value of 275.2W.

Remarks on Future Developments

As we have seen in this section, optical device technologies have made significant progress in the 1990s and reached a level that warrants their practical application to large-scale optical path cross-connect systems. However, further development is necessary to improve their performance and to attain more cost-effective implementation of optical devices/components within a card in transport system cabinets, which can be realized by fully utilizing optical interconnection technologies.

To what extent optical technologies will be applied to telecommunications networks, or to what extent the massive potential of optical technologies can be realized beyond the point-to-point transmission application, depends on the advances in optical device/component technology. The optical path technology developments discussed in this chapter are important in that they cast a light on the right way of advancing optical devices/components, and so minimize the time for worldwide deployment of optical networking technologies.

4.7 OPTICAL PATH ECONOMICS

4.7.1 Existing Transport Network Cost

This section provides an example of the cost reduction possible with optical path technologies. First, the cost of an SDH-based transport network is explained using Japan's middle- and long-haul trunk transmission networks as an example [157]. Figure 4.45(a) illustrates the model used in the evaluation, and the extent of the cost evaluation. End-to-end VC-4 path cost was evaluated for interoffice transmission system speeds ranging from 600 Mbps to 10 Gbps. The network parameters used for the evaluation are listed below.

- Average end-to-end path length is 200 km;
- One cross-connection and one regeneration are done for each path;
- Intraoffice interface speed of 156 Mbps.

Figure 4.46 shows the cost evaluation results and details. Note that repeater cost is included in the node cost (not in the line cost), and network OA&M cost is not included. Link utilization was assumed to be 100%, so line cost is inversely proportional to the line bit rate. As shown in Figure 4.46, when the

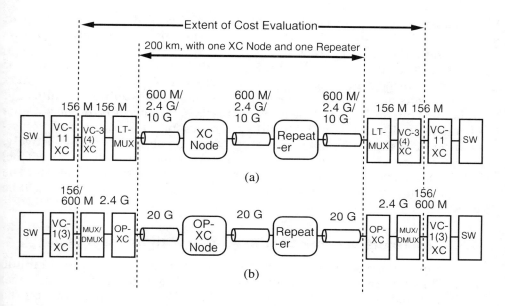

Figure 4.45 Transport cost evaluation model—long- and middle-distance trunk line network model. (*Source:* [157]. Reprinted with permission.)

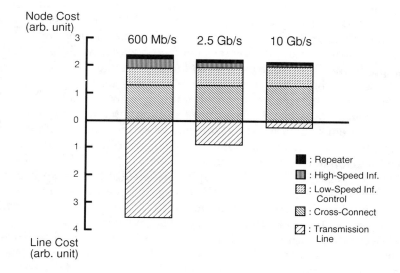

Figure 4.46 SDH network cost (electrical path network). (*Source:* [157]. Reprinted with permission.)

transmission speed exceeds 2.5 Gbps, the major cost of the end-to-end path connection is the node cost, which is dominated by cross-connect and low speed intraoffice interface costs.

When enhancing transport network bandwidth towards B-ISDN, it is clear that total transport network cost can be effectively reduced by node cost reduction. This is possible by increasing cross-connection unit-bandwidth and intraoffice interface speed. The former is reasonable, since it is in line with the channel speed increase in the B-ISDN environment. This will realize a cross connect throughput of more than several hundreds of gigabits cost-effectively

Enhancement of interoffice transmission speed is necessary, of course since line cost per bit is inversely proportional to the transmission bit rate. The next section will show how optical path technologies exploiting WDM will be effective in attaining this.

4.7.2 Transport Network Cost Reduction With Optical Paths

To eliminate the electrical processing bottleneck at transport nodes, all traffic except that terminating at the node will be cross-connected at the optical level with wavelength routing of optical paths [154]. Node cost reductions can be achieved since termination of the time division multiplexed total line capacity is not required and the cross-connection for trunk access can be done at higher levels more easily. With WDM, wavelength paths are not synchronized to each other, so modular growth capability of the cross-connect system can be more easily achieved than with TDM, as is demonstrated by the cross-connect architecture described in Section 4.6. Again, this is a very important point for the economical introduction of a cross-connect system with potentially vast throughput to transport nodes whose throughput demands range from small to large. Since broadband needs may emerge abruptly in diverse areas, it will be very effective to achieve cross-connect modular growth capability and flexible and cost-effective broadband service provisioning capability without additional outdoor plant construction, which minimizes service introduction delay.

Figure 4.47 shows an estimate of the reduction in transport network cost achieved with optical paths as evaluated for the VC-4 paths. The model network used for the evaluation is depicted in Figure 4.45(b) and is equivalent to the existing SDH network in Figure 4.45(a). The optical path cross-connect architecture utilized for this cost evaluation is the DC-switch-based architecture explained in Section 4.6. Considering available cost-effective optical techniques, the optical path bandwidth is set to 2.5 Gbps [144]. The cross-connect is assumed to have 16 input/output ports and 8 wavelengths per fiber, so the total main frame throughput is 320 Gbps. The experimental performance of this cross-connect system has been confirmed [160,157]. Cross-connect system throughput can be enhanced to 900 Gbps by utilizing three main frames an

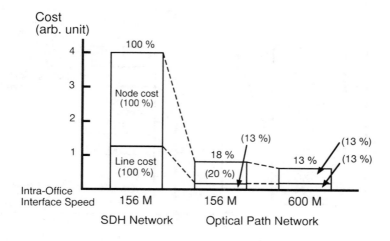

Figure 4.47 Transport network cost reduction with optical path technologies. (*Source:* [157]. Reprinted with permission.)

two junction frames [166]. With optical path technology, the end-to-end VC-4 path cost can be reduced (see Figure 4.47) to 18% and 13% of that realized with existing SDH systems utilizing 2.5 Gbps interoffice transmission, for intraoffice interface speeds of 156 Mbps and 600 Mbps, respectively (see Figure 4.45(b)). The line cost reduction stems from the total transmission speed increase achieved with WDM, from 2.5 Gbps (TDM) to 20 Gbps (2.5 Gbps × 8 WDM). The node cost reduction achieved by the adopted architecture is based on the increased cross-connect unit of 2.5 Gbps (optical path bandwidth is 2.5 Gbps) for trunk access and restoration and increased intraoffice interface speed of 2.5 Gbps to the optical path cross-connect while the interface speed between VC-3(1) XT and LT-Mux is assumed to be 156 Mbps or 600 Mbps (see Figure 4.45(b)). It should be noted that this architectural change (increasing cross-connection unit bandwidth, intraoffice interface speed, and interoffice transmission speed) can be functionally possible with enhanced electrical technologies and higher order 2.5 Gbps electrical paths; however, no very effective solution has emerged to realize such a large-throughput electrical cross-connect system within one cabinet. In contrast, optical technologies that enable this already exist, as explained thus far. Furthermore, optical path technologies provide other benefits that cannot be achieved with electrical technologies, as discussed in Section 4.4.

Details of the evaluated optical path cross-connect system cost are shown in Figure 4.48. It should be noted that more than 50% of the cross-connect cost comes from the electrical system of the MUX/DMUX and intraoffice interface

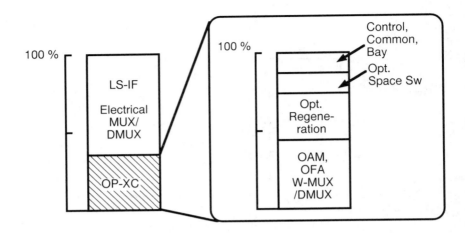

Figure 4.48 Details of optical path cross-connect system cost. (*Source:* [157]. Reprinted with permission.)

cost. Optical regeneration cost (electrical regeneration is used instead of linear repeaters utilizing optical-fiber amplifiers (OFAs)) makes up about 12% of the total node cost, which may not be so significant when its effectiveness, discussed in Section 4.6, is considered.

4.8 CONCLUSION

Our transport network has evolved over the last 15 years with the introduction of optical technologies, as reviewed in this chapter. The kinds of optical components actually implemented in the network, however, have been rather limited-optical fibers and connectors, laser diodes and light emitting diodes, photodiodes, optical splitters, recently introduced OFAs, and so on. The optical components that will be widely introduced in the next stage to our networks and that will enhance the application extent of optical network technologies will be wavelength multiplexers and demultiplexers, optical space division switches and wavelength-tunable optical sources and filters. Based on these advanced optical component technologies, we will be closer to creating future bandwidth-abundant and cost-effective networks.

We discussed in this chapter an exciting new network technology: optical paths. Different optical path realization technologies were discussed and compared; they will be utilized in global area networks, MANs, and LANs. Depending on the application, the optical benefits that should be exploited, which will differ, need to be emphasized in order to justify the introduction.

For example, for trunk transport network applications, large-capacity and long-haul transmission capability are the first and major benefits and justified the introduction of optical technologies to our networks for point-to-point transmission. These benefits need to be maximally utilized in the next step in conjunction with the new functionality of wavelength routing, which enables cost-effective traffic routing (especially for broadband applications). Thus, by combining the WDM transmission and wavelength-routing capabilities, we realize much wider application areas in the not-so-distant future. This chapter discussed some of the key technologies necessary to enable this.

On the other hand, for LAN applications, transmission distance is usually very limited, so other optical attributes need to be emphasized such as easy realization of broadcast-type communication capability and the transparency of optical channels/paths to allow the technology to penetrate application areas that are extremely cost-sensitive.

Application of optical technologies to call-by-call related functions, call-by-call wavelength routing in a public telecommunications network, and photonic switching will be seen only after the optical processing technologies have matured and their costs reduced.

Grasping the right direction to proceed and promoting the development of the necessary optical technologies are very important tasks of telecommunications network engineers. The development and standardization of necessary OA&M technologies, not discussed in this chapter, are also critical to the rapid and wide deployment of photonic technologies discussed in this chapter.

References

[1] Kao, K. C., and G. A. Hockham, "Dielectric-fiber surface waveguides for optical frequencies," *Proc. IEEE*, Vol. 113, 1966, p. 1151.
[2] Kapron, F. P., D. B. Keck, and R. D. Maurer, "Radiation losses in glass optical waveguides," *Appl. Phys. Letters*, Vol. 17, 1970, pp. 423–425.
[3] Miya, T., Y. Terunuma, T. Hosaka, and T. Miyashita, "Ultimate low-loss single mode fiber at 1.55 μm," *Electron. Letters* Vol. 15, 1976, p. 106.
[4] Nathan, M., W. Dumke, G. Burns, F. Dill, and G. Lasher, "Stimulated emission of radiation from GaAs p-n junctions," *Appl. Phys. Letters*, Vol. 1, 1962, p. 62.
[5] Hayashi, I., M. B. Panishi, P. W. Foy, and S. Sumski, "Junction lasers which operate continuously at room temperature," *Appl. Phys. Letters* Vol. 17, 1970, p. 109.
[6] Shimada, S., and N. Uchida, "Field trial of medium/small capacity optical fiber transmission systems," *ECL Technical Journal*, NTT, Japan, Vol. 30, No. 9, 1981, pp. 2121–2132.
[7] *IEEE Communications Magazine*, Special Issue on Fiber-optic subscriber loops, Vol. 32, No. 2, Feb. 1994.
[8] Ross, F. E., "FDDI - A tutorial," *IEEE Communications Magazine*, Vol. 24, No. 5, May 1986.
[9] Getchell, D., and P. Rupert, "Fiber channel in the local area network," *IEEE LTS*, Vol. 3, No. 2, May 1992, pp. 38–42.
[10] ANSI X3T9/91-062, X3T9.3/92-007, Rev 2.2, Fiber Channel - Physical and Signaling Interface, Jan. 1992.

[11] Tada, K., and H. S. Hinton (Eds.), *Photonic Switching II*, Springer-Verlag, 1990.
[12] Hinton, H. S., and J. E. Midwinter (Eds.), *Photonic Switching*, IEEE Press, 1990.
[13] *IEEE Communications Magazine*, Photonic Switching, Vol. 25, No. 5, May 1987.
[14] Nakagawa, K., K. Aida, K. Aoyama, and K. Hohkawa, "Optical amplification in trunk transmission networks," *IEEE LTS*, Vol. 3, No. 1, Feb. 1992, pp. 19–26.
[15] Nakagawa, K., K. Hagimoto, K. Nishi, and K. Aoyama, "A bit-rate flexible transmission field trial over 300-km installed cables employing optical fiber amplifiers," *OAA '91*, PD-12, July 1991.
[16] Kobayashi, Y., Y. Sato, K. Aida, K. Hagimoto, and K. Nakagawa, "SDH-based 10 Gbit/s optical transmission system," *GLOBECOM '94*, San Francisco, CA, Dec. 1994, pp. 1166–1170.
[17] *IEEE Communications Magazine*, Feature Topic -SDH/SONET-, Vol. 31, No. 9, Sept. 1993.
[18] Shirakawa, H., K. Maki, and H. Miura, "Japan's network evolution relies on SDH-based systems," *IEEE LTS*, Nov. 1991, pp. 14–18.
[19] Wu, T-H., *Fiber Network Service Survivability*, Norwood, MA: Artech House, 1992.
[20] ITU-T Recommendation I.233.1, ISDN frame relaying bearer service, Oct. 1991.
[21] ITU-T Recommendation I.555, Frame relaying bearer service interworking, July 1993.
[22] Bellcore Technical Information, TA-TSV-001059, Generic Requirements for SMDS Networking, Aug. 1992.
[23] Bellcore Technical Information, TA-TSV-001064, Generic Criteria on Operations Interfaces - SMDS Information Model and Usage, Dec. 1992.
[24] Armbruster, H., "The flexibility of ATM: Supporting future multimedia and mobile communications," *IEEE Communications Magazine*, Feb. 1995, pp. 76–84.
[25] *IEEE Communications Magazine*, Special Issue of Data Communications in ATM Networks, Vol. 32. No. 8, Aug. 1994.
[26] *IEEE Communications Magazine*, Special Issue on Fiber-Optic Subscriber Loops, Vol. 32, No. 2, Feb. 1994.
[27] Miki, T., "Fiber-optic access networks and services," *NTT Review*, Vol. 6, No. 3, May 1994, pp. 17–25.
[28] Ogawa, K., "Access network systems," *NTT Review*, Vol. 6, No. 4, July 1994, pp. 25–33.
[29] ISO/IEC 11172-1:1993 Information Technology - Coding of moving pictures and associated audio for digital storage media at up to about 1.5 Mbit/s - Part 1: Systems.
[30] ISO/IEC 11172-2:1993 Information Technology - Coding of moving pictures and associated audio for digital storage media at up to about 1.5 Mbit/s - Part 2: Video.
[31] ISO/IEC 13818-2:1993, Recommendation H. 262, Generic coding of moving pictures and associated audio.
[32] Hirota, Ken'ichiro, "Future of video communication," *Weekly Telecom*, March 13, 1986, pp. 1–4.
[33] Sato, K., "Transport network evolution with optical paths," *Proc. ECOC '94*, Florence, Italy, Sept. 25–29, 1994, pp. 919–926.
[34] Ishio, H., "Progress of fiber optic technologies and their impact on future telecommunication networks," *Proc. FORUM '91* (Technical Symposium of TELECOM 91), Geneva, Switzerland, Oct. 10–15, 1991, Session 2.2, pp. 117–120.
[35] Cheung, Nim K., "The infrastructure for gigabit computer networks," *IEEE Communications Magazine*, April 1992, pp. 60–68.
[36] Emmanuel, Desurvire, *Erbium-Doped Fiber Amplifiers*, New York, NY: John Wiley & Sons, Inc., 1994.
[37] Bjarklev, Anders, *Optical Fiber Amplifiers: Design and System Application*, Norwood, MA: Artech House, 1993.
[38] Hasegawa, A., and F. Tappert, "Transmission of stationary nonlinear optical pulses in disper-

sive dielectric fibers I. Anomalous dispersion," *Appl. Phys. Lett.*, Vol. 23, No. 3, 1973, pp. 142–144.
[39] Hasegawa, A., *Optical Soliton in Fibers*, Berlin, Germany: Springer-Verlag, 1989.
[40] Agrawal, G. P., *Nonlinear Fiber Optics*, San Diego, CA: Academic Press Inc., 1989.
[41] *IEEE J. Select. Areas Commun.*, Special Issue on Dense wavelength division multiplexing techniques for high capacity and multiple access communication systems, Vol. 8, No. 6, Aug. 1990.
[42] Goodman, M. S., "Multiwavelength networks and new approaches to packet switching," *IEEE Commun. Magazine*, Vol. 27, No. 10, Oct. 1989, pp. 27–35.
[43] Wagner, S. S., and H. Kobrinski, "WDM applications in broadband telecommunication networks," *IEEE Communications Magazine*, March 1989, pp. 22–30.
[44] Mukherjee, B., "WDM-based local lightwave networks Part I: Single-hop systems" *IEEE Networks*, May 1992, pp. 12–28.
[45] *IEEE Networks*, Special Issue on Optical Multiaccess Networks, Vol. 3, No. 2, March 1989.
[46] Toba, H., K. Oda, K. Nakanishi, N. Shibata, K. Nosu, N. Takato, and M. Fukuda, "A 100-channel optical FDM transmission/distribution at 622 Mb/s over 50 km," *J. Lightwave Technology*, Vol. 8, No. 9, Sept. 1990, pp. 1396–1401.
[47] Saruwatari, M., "High-speed optical signal processing for communications systems," *IEICE Trans. Commun.*, Vol. E78-B, No. 5, May 1995, pp. 635–643.
[48] Kawanishi, S., T. Morioka, O. Kamatani, H. Takara, and M. Saruwatari, "Time-division-multiplexed 100 Gbit/s, 200 km optical transmission experiment using PLL timing extraction and all optical demultiplexing based on polarization insensitive four-wave-mixing," *Tech. Digest of OFC '94*, PD23, 1994, pp. 108–111.
[49] Miura, T., H. Nishihata, and T. Yamashita, "Versatile digital system switcher," *NTT Shisetsu*, Vol. 38, No. 5, May 1986, pp. 71–77 (in Japanese).
[50] Shirakawa, H., K. Maki, and H. Miura, "Japan's network evolution relies on SDH-based systems," *IEEE LTS*, Nov. 1991, pp. 14–18.
[51] Nishihata, K., "Trunk link system," *NTT Review*, Vol. 6, No. 6, Nov. 1994, pp. 24–32.
[52] CCITT I-series Recommendations (B-ISDN), 1992.
[53] Obara, H., I. Tokizawa, and N. Tokura, "ATM cross-connect system for broadband ISDN," *Proc. FORUM '91*, Geneva, Switzerland, Oct. 10–15, 1991, Session 2.5, pp. 265–269.
[54] Eng, K. Y., M. J. Karol, G. J. Cyr, and M. A. Pashan, "A 160-Gb/s ATM switch prototype using the concentrator-based growable switch architecture," *Proc. ICC '95*, Seattle, WA, June 18–22, 1995, pp. 550–554.
[55] Eng, K. Y., and M. J. Karol, "State of the art in gigabit ATM switching," *Proc. Broadband Switching Symposium '95*, Poznan, Poland, April 19–21, 1995, pp. 3–20.
[56] Hadama, H., and S. Ohta, "Routing control of virtual paths in large-scale ATM-based transport network," *Trans. on IEICE Japan*, Vol. J72-B-I, No. 11, Nov. 1989, pp. 970–978 (in Japanese).
[57] Kawamura, R., K. Sato, and I. Tokizawa, "Self-healing network techniques utilizing Virtual Paths," *5th Int. Network Planning Symposium, Networks '92*, Kobe, Japan, May 17–22, 1992, pp. 129–134.
[58] Kawamura, R., K. Sato, and I. Tokizawa, "Self-healing ATM networks based of virtual path concept," *IEEE J. Selected Areas Commun.*, Vol. 12, No. 1, Jan. 1994, pp. 120–127.
[59] Wu, T. H., "Emerging technologies for fiber network survivability," *IEEE Communications Magazine*, Feb. 1995, pp. 58–74.
[60] Multimedia Planning and Promotion Office, NTT, "Joint utilization tests of multimedia communications," *NTT Review*, Vol. 6, No. 4, July 1994, pp. 8–10.
[61] Shimizu, A., and H. Uose, "NTT's strategy for high speed multimedia communications," *Proc. INET '94*, Czecho, June. 15–17, 1994, pp. 253-1–253-8.

[62] Falconer, W. E., "Service assurance in modern telecommunications networks," *IEEE Communications Magazine*, Vol. 28, No. 6, June 1990, pp. 32–39.
[63] Wu, T-H., *Fiber Network Service Survivability*, Norwood, MA: Artech House, 1992.
[64] *IEEE J. Selected Areas Commun.*, Special Issue on Integrity of Public Telecommunication Networks, Vol. 12, No. 1, 1994.
[65] Grover, W. D., "The self-healing network," *Proc. GLOBECOM '87*, Tokyo, Japan, Nov. 15–18 1987, pp. 28.2.1–28.2.6.
[66] Hasegawa, S., A. Kanemasa, H. Sakaguchi, and R. Maruta, "Dynamic reconfiguration of digital cross-connect systems with network control and management," *Proc. GLOBECOM '87*, Tokyo Japan, Nov. 15–18, 1987, pp. 28.3.1–28.3.5.
[67] Yang, C. H., and S. Hasegawa, "FITNESS-Failure immunization technology for network service survivability," *Proc. GLOBECOM '88*, Nov. 28–Dec. 1, Lauderdale, FL pp. 47.3.1.–47.3.6.
[68] Sato, K., and S. Okamoto, "Evolution of optical path layer techniques toward photonic networks," *Proc. of the 1992 IEICE Fall Conference*, Sept. 27–30, Tokyo, Japan, SB-7-1.
[69] Okamoto, S., and K. Sato, "Transport network architecture using optical paths," *Proc. of the 1992 IEICE Fall Conference*, Sept. 27–30, Tokyo, Japan, B-505.
[70] Okamoto, S., and K. Sato, "Photonic transport network and the cross-connect system architecture," *Technical Report of IEICE*, CS92-49, Oct. 22–23, 1992.
[71] Sato, K., S. Okamoto, and H. Hadama, "Optical path layer technologies to enhance B-ISDN performance," *Proc. ICC '93*, Geneva, Switzerland, May 23–26, 1993, pp. 1300–1307.
[72] Sato, K., S. Okamoto, and H. Hadama, "Network reliability enhancement with optical path layer technologies," *IEEE J-SAC*, Vol. 12, No. 1, Jan. 1994, pp. 159–170.
[73] Kobrinski, H., "Crossconnection of wavelength-division-multiplexed high-speed channels," *Electronics Letters*, Vol. 23, No. 18, Aug. 1987, pp. 974–976.
[74] Hill, G., "Wavelength routing approach to optical communication networks," *Br. Telecom Technical J.*, Vol. 6, No. 3, July 1988, pp. 24–30.
[75] Hill, G., P. J. Chidgy, and J. Davidson, "Wavelength routing for long haul networks," *Proc. ICC '89*, Boston, MA, June 11–14, 1989, pp. 23.3.1.–23.3.5.
[76] Hill G., I. Hawker and P. J. Chidgy, "Application of wavelength routing in a core telecommunications," *IEE Int. Conf. on Integrated Broadband Services and Networks*, London, England, Oct. 15–18, 1990, pp. 285–289.
[77] ITU-T Recommendation I.432, B-ISDN user-network interface -physical layer specification, March 1993.
[78] ITU-T Recommendation I.311, B-ISDN general network aspects, March 1993.
[79] Okamoto, S., K. Oguchi, and K. Sato, "Network architecture and management concepts for optical transport networks," *IEEE NOMS '96*, Kyoto, Japan, April 15–19, 1996.
[80] Ueda, H., H. Tsuji, and T. Tsuboi, "New synchronous digital transmission system with network node interface," *Proc. GLOBECOM '89*, Dallas, TX, Nov. 27–30, 1989, pp. 42.4.1–42.4.5.
[81] Uematsu, H., and H. Ueda, "STM signal transfer techniques in ATM networks," Proc. ICC '92, Chicago, IL, June 14–18, 1992, pp. 311.6.1–311.6.5.
[82] Okamoto, S., N. Nagatsu, K. Oguchi, and K. Sato, "Management concept of optical path network," *First IEEE COMSOC Workshop on WDM Optical Network Management and Control*, Seattle, WA, June 18, 1995.
[83] Hadama, H., and S. Ohta, "Routing control of virtual paths in large-scale ATM-based transport network," *Trans. IEICE Japan*, Vol. J72-B-I, No. 11, Nov. 1989, pp. 970–978 (in Japanese).
[84] Dono, N. R., P. E. Green, Jr., K. Liu, R. Ramaswami, and F. F. Tong, "A wavelength division multiple access network for computer communication," *IEEE J. Select. Areas Commun.*, Vol. 8, No. 6, Aug. 1990, pp. 983–994.

[85] Mukherjee, B., "WDM-based local lightwave networks Part I: Single-hop systems" *IEEE Networks*, May 1992, pp. 12–28.
[86] Sivaraajan, K. N., "Multihop logical topologies for gigabit lightwave networks," *IEEE LTS*, Aug. 1992, pp. 20–25.
[87] Mukherjee, B., "WDM-based local lightwave networks Part II: multihop systems," *IEEE Networks*, July 1992, pp. 20–31.
[88] Marsan, M. A., A. Binco, E. Leonardi, and F. Neri, "Topologies for wavelength-routing all-optical networks," *IEEE/ACM Trans. on Networking*, Vol. 1, No. 5, Oct. 1993, pp. 534–546.
[89] Li, B., and A. Ganz, "Virtual topologies for WDM star LANs -The regular structure approach," *INFOCOM '92*, 1992, pp. 9B.3.1–9B.3.10.
[90] Marsan, M. A., A. Bianco, E. Leonardi, and F. Neri, "Topologies for wavelength-routing all optical networks," *IEEE/ACM Trans. on Networking*, Vol. 1, No. 5, Oct. 1993, pp. 534–546.
[91] Karol, M., and B. Glance, "Performance of the PAC optical packet network," *GLOBECOM '91*, 1991, pp. 1258–1263.
[92] Lee, T. T., M. S. Goodman, and E. Arthurus, "A broadband optical multicast switch," *ISS '90*, Stockholm, Sweden, May 27–June 1, 1990, Proc. Vol. III, pp. 7–13.
[93] Arthurs, E., J. M. Cooper, M. S. Goodman, H. Kobrinski, M. Tur, and M. P. Vecchi, "Multiwavelength optical crossconnect for parallel-processing computers," *Electron. Letters*, Vol. 24, No. 2, 1988, pp. 119–120.
[94] Acampora, A. S., M. J. Karol, and M. G. Hluchyj, "Terabit lightwave networks: the multihop approach," *AT&T Technical Journal*, Vol. 66, Issue 6, Nov./Dec. 1987, pp. 21–34.
[95] Karol, M. J., and S. Z. Shaikh, "A simple adaptive routing scheme for congestion control in shuffflenet multihop lightwave networks," *IEEE Journal on Selected Areas in Communications*, Vol. 9, No. 7, Sept. 1991, pp. 1040–1051.
[96] Zhang, Z., and A. S. Acampora, "Analysis of all optical multihop networks," *GLOBECOM '90*, 1990, pp. 903.7.1–903.7.7.
[97] Banerjee, S., and D. Sarkar, "Hypercube Connected Rings: A fault-tolerant and scalable architecture for virtual lightwave network topology," *Proc. INFOCOM '94*, 1994, pp. 9d.1.1.–9d.1.8.
[98] Eng, K. Y., "A multi-fiber ring architecture for distributed lightwave networks," *Proc. ICC '88*, Philadelphia, PA, June 12–15, 1988, pp. 46.3.1–46.3.7.
[99] Chlamtac, I., A. Ganz, and G. Karmi, "Lightnet: Lightpath based solutions for wide bandwidth WANs," *Proc. INFOCOM '90*, California, June 3–4, 1990, pp. 1014–1021.
[100] Maxemchuk, N. F., "Regular mesh topologies in local and metropolitan area network," *AT&T Technical Journal*, Vol. 64, No. 7, Sept. 1985, pp. 1659–1685.
[101] Sen, A., and P. Maitra, "Comparative study of Shuffle-exchange, Manhattan street and Supercube network for lightwave applications," *Proc. GLOBECOM '91*, 1991, pp. 1849–1854.
[102] Sivarajan, K., and R. Ramaswami, "Multihop lightwave networks based on de bruijn graphs," *INFOCOM '91*, 1991, pp. 9A.3.1–9A.3.11.
[103] Sivarajan, K., and R. Ramaswami, "Lightwave networks based on de bruijn graphs," *IEEE/ACM Trans. on Networking*, Vol. 2, No. 1, Feb. 1994, pp. 70–79.
[104] Banister, J. A., L. Fratta, and M. Gerla, "Topological design of the wavelength-division optical network," *Proc. INFOCOM '90*, California, June 3–4, 1990, pp. 1005–1013.
[105] Gerla, M., M. Kovacevic, and J. A. Banister, "Multilevel optical networks," *Proc. ICC '92*, Chicago, June 14–18, 1992, pp. 340.4.1–340.4.5.
[106] Chlamtac, I., A. Ganz, and G. Karmi, "Purely optical networks for terabit communication," *Proc. INFOCOM '89*, Ottawa, April 23–24, 1989, pp. 887–896.
[107] Bala, K., T. E. Stern, and K. Bala, "Algorithms for routing in a linear lightwave network," *Proc. INFOCOM '91*, FL, April 9-11, 1991, pp. 2A.1.1.–2A.1.9.

[108] Goodman, M., H. Kobrinski, and K. W. Lob, "Application of wavelength division multiplexing to communication network architectures," *ICC 86*, Toronto, June 22–25, 1986, pp. 29.4.1–29.4.3.
[109] Goodman, M., C. Brackett, R. Bulley, C. Lo, H. Kobrinski, and M. Vecchi, "Design and demonstration of the LAMBDANET™ System: A multiwavelength optical network," *GLOBECOM' 87*, Tokyo, Nov. 15–18, 1987, pp. 37.4.1–37.4.4.
[110] Kobrinski, H., "Crossconnection of wavelength-division-multiplexed high-speed channels," *Electronics Letters*, Vol. 23, No. 18, Aug. 27th, 1987, pp. 974–976.
[111] Hill, G. R., "A wavelength routing approach to optical communication networks," *Br. Telecom Technical Journal*, Vol. 6, No. 3, July 1988, pp. 24–31.
[112] Okano, Y., T. Kawata, and T. Miki, "Designing digital paths in transmission networks," *Proc. GLOBECOM '86*, Houston, TX, 1986, pp. 25.2.1–25.2.5.
[113] Okano, Y., S. Ohta, and T. Kawata, "Assessment of cross-connect systems in transmission networks," *Proc. GLOBECOM '87*, Tokyo, Japan, Nov. 15–18, 1987.
[114] Sato, K., and I. Tokizawa, "Flexible asynchronous transfer mode networks utilizing Virtual Paths," *Proc. ICC '90*, Atlanta, GA, April 16–19, 1990, pp. 318.4.1.–318.4.8.
[115] Chlamtac, I., A. Ganz, and G. Karmi, "An approach to high bandwidth optical WAN's," *IEEE Trans. Commun.*, Vol. 40, No. 7, July 1992, pp. 1171–1182.
[116] Nagatsu, N., Y. Hamazumi, and K. Sato, ""Number of wavelengths required for constructing large scale optical path networks," *Trans. on IEICE Japan*, Vol. J77-B-I, No. 12, Dec. 1994, pp. 728–737.
[117] Nagatsu, N., Y. Hamazumi, and K. Sato, "Optical Path Accommodation Designs Applicable to Large Scale Networks," *Trans. on IEICE Japan*, Vol. E78-B, No. 4, April 1995, pp. 597–607.
[118] Nagatsu, N., Y. Hamazumi, and K. Sato, "Optical path Accommodation Design," *Technical Report of IEICE Japan*, Nov. 1993, pp. CS93-137.
[119] Wauters, N., and P. Demeester, "Wavelength requirements and survivability in WDM cross-connected networks," *Proc. ECOC '94*, Firenze, Italy, Sept. 25–29, 1994.
[120] Lee, K. -C., and V.O.K. Ki, "A wavelength-convertible optical network," *IEEE J. Lightwave Technology*, Vol. 11, No. 5/6, May/June 1993, pp. 962–970.
[121] Ramaswami, R., and K. N. Sivarajan, "Optimal routing and wavelength assignment in all-optical networks," *Proc. INFOCOM '94*, Toronto, Canada, June 14–16, 1994, pp. 970–679.
[122] Zhang, Z., and A. Acampora, "Heuristic Wavelength assignment algorithm for multihop WDM networks with wavelength routing and wavelength reuse," *Proc. INFOCOM '94*, Toronto, Canada, June 14–16, 1994, pp. 534–543.
[123] Chan, K., and T-S. P. Yum, "Analysis of least congested path routing in WDM lightwave networks," *Proc. INFOCOM '94*, Toronto, Canada, June 14–16, 1994, pp. 962–969.
[124] Hamazumi, Y., N. Nagatsu, S. Okamoto, and K. Sato, "Number of wavelengths required for constructing optical path network considering failure restoration," *Trans. on IEICE Japan*, Vol. J77-B-I, No. 5, 1994, pp. 275–284.
[125] Hamazumi, Y., N. Nagatsu, and K. Sato, "Number of Wavelengths Required for Optical Networks Considering Failure Restoration," *Proc. OFC '94*, San Jose, CA, Feb. 21–26, 1994, TuE2.
[126] ITU-T Revised Text for Draft Rec. G.lon (version I), COM 15-276-E, Geneva, Switzerland, Nov. 13–24, 1995.
[127] ITU-T Revised Draft of Rec. G.mcs, Temporary Document 45 (4/15), Annex 4, Geneva, Switzerland, March 1996.
[128] Nagatsu, N., S. Okamoto, and K. Sato, "Optical path accommodation considering failure restoration with minimum cross-connect system scale," *Proc. NOMS '96*, Kyoto, Japan, April 15–19, 1996, 10.1, pp. 213–224.
[129] Nagatsu, N., and K. Sato, "Optical path accommodation design enabling cross-connect system

scale evaluation," *IEICE Trans. on Communications*, Vol. E78-B, No. 9, Sept. 1995, pp. 1339–1343.
[130] Nagatsu, N., S. Okamoto, and K. Sato, "Optical path cross-connect system evaluation using path accommodation design for restricted wavelength multiplexing," Special Joint Issue of *IEEE Journal on Selected Areas in Communications and IEEE Journal of Lightwave Technology*, Vol. 14, No. 5, June 1996.
[131] Byer, R. L., and R. L. Herbst, *Topics in Applied Physics*, Vol. 16, Springer-Verlag, 1977.
[132] Murata, S., A. Tomita, J. Shimizu, M. Kitayama, and S. Suzuki, "Observation of highly nondegenerate four-wave mixing (>1 THz) in an InGaAsP multiple quantum well laser," *Appl. Phys. Letter*, Vol. 58, No. 8, April 1991, pp. 1458–1460.
[133] Iannone, E., and R. Sabella, "Performance evaluation of an optical multi-carrier networking using wavelength converters based on FWM in semiconductor optical amplifiers," *JLT*, Vol. 13, No. 2, Feb. 1995, pp. 312–324.
[134] Inoue, K., and H. Toba, "Wavelength conversion experiment using fiber four-wave mixing," *IEEE Photonics Technology Letters*, Vol. 4, No. 1, Jan. 1992, pp. 69–72.
[135] Inoue, K., T. Hasegawa, K. Oda, and H. Toba, "Multichannel frequency conversion experiment using fiber four-wave mixing," *IEEE Electronics Letters*, Vol. 29, No. 19, Sept. 1993, pp. 1708–1709.
[136] Jopson, R. M., and R. E. Tench, "Polarisation-independent phase conjugation of lightwave signals," *IEEE Electronics Letters*, Vol. 29, No. 25, Dec. 1993, pp. 2216–2217.
[137] Odagiri, Y., K. Komatsu, and S. Suzuki, "Bistable laser diode memory for optical time-division switching applications," *Proc. CLEO '84*, Anaheim, CA, 1984.
[138] Hata, S., M. Ikeda, Y. Noguchi, and S. Kondo, "Monolithic integration of an InGaAs PIN photodiode, two InGaAs column gate FETs and InGaAsP laser for optical regeneration," *Proc. 17th Conf. Solid State Devices and Materials*, Tokyo, Japan, 1985, pp. 79–82.
[139] Sharony, J., T. E. Stern, and K. W. Cheung, "The wavelength dilation concept -Implementation and system considerations," *Proc. ICC '92*, Chicago, IL, June 14–18, 1992, pp. 330.3.1–330.3.8.
[140] O'Mahony, M. J., "Transmission design aspects of multi-wavelength optical networks," *ICC '93 Workshop on Challenges of All Optical Networks*, Geneva, Switzerland, May 27th, 1993, Paper 8, pp. 55–60.
[141] Saxtoft, C., and P. Chidgey, "Error rate degradation due to switch crosstalk in large modular switched optical networks," *IEEE Photonics Technology Letters*, Vol. 5, No. 7, July 1993, pp. 828–831.
[142] Zhou, J., M. J. O'Mahony, R. Caadeddu, and E. Casaccia, "Cross-talks in multiwavelength optical cross-connect networks," *Proc. OFC '95*, San Diego, February 26–March 3, 1995, pp. 278–280.
[143] Testa, F., S. Merle, and P. Pagnan, "Fabry-perott bandwidth effects on end to end transmission performances in a multiwavelength transport network," *IEEE Photonics Technology Letters*, Vol. 6, No. 8, July 1994, pp. 1027–1030.
[144] Okamoto, S., A. Watanabe, and K. Sato, "Optical path cross-connect architecture for photonic transport network," Special Joint Issue of *IEEE Journal of Lightwave Technology and IEEE Journal on Selected Areas in Communications*, Vol. 14, No. 6, 1996.
[145] Okamoto, S., and K. Sato, "Optical path cross-connect systems for photonic transport network," *Proc. GLOBECOM '93*, Nov. 29–Dec. 2, 1993, Houston, TX, pp. 474–480.
[146] Clos, C., "A study of non-blocking switching networks," *Bell System Technical Journal*, Vol. 32, No. 2, 1953, pp. 406–424.
[147] Watanabe, A., S. Okamoto, and K. Sato, "Optical Path Cross-Connect Node Architecture Offering High Modularity for Virtual Wavelength Paths," *Trans. on IEICE Japan*, Vol. E78-B, No. 5, May 1995, pp. 686–693.
[148] Hill, G. R., et al., "A transport network layer based on optical network elements," *J. Lightwave Technology*, Vol. 11, No. 5/6, May/June 1993, pp. 667–679.

[149] Chidgey, P. J., G. R. Hill, and C. Saxtoft, "Wavelength and space switched optical networks and nodes," *Proc. ISS '92*, Yokohama, Japan, Oct. 25–30, 1992, pp. 352–356.

[150] Smith, D. A., J. J. Johnson, J. E. Baran, K. W. Cheung, and S. C. Liew, "Integrated acoustically tunable optical filters: devices and applications," *Proc. OFC '79*, 1991, p. 142.

[151] Nishio, M., and S. Suzuki, "Photonic wavelength-division switching network using parallel λ-switch," *Proc. PS '90*, PD14B-9, 1990.

[152] Okamoto, S., A. Watanabe, and K. Sato, "A new optical path cross-connect system architecture utilizing delivery and coupling matrix switch," *IEICE Trans. Commun., Japan*, Vol. E77-B, No. 10, Oct.. 1994, pp. 1272–1274.

[153] Watanabe, A., S. Okamoto, and K. Sato, "Optical Path Cross-Connect Node Architecture with High Modularity for Photonic Transport Network," *IEICE Trans. Commun., Japan*, Vol. E77-B, No. 10, Oct. 1994, pp. 1220–1229.

[154] Sato, K., S. Okamoto, and A. Watanabe, "Optical paths and realization technologies," *Proc. GLOBECOM '94*, San Francisco, CA, Nov. 28–Dec. 2, 1994, pp. 1513–1520.

[155] Spanke, R. A., "Architectures for guided-wave optical space switching systems," *IEEE Communications Magazine*, Vol. 25, No. 5, May 1987, pp. 42–48.

[156] Nagase, R., A. Himeno, K. Kato, and M. Okuno, "Silica-based 8x8 optical-matrix switch module with hybrid integrated driving circuits," *Proc. ECOC '93*, Sept. 12–16, 1993, Montreux, Switzerland, MoP1.2.

[157] Sato, K., S. Okamoto, and A. Watanabe, "Photonic transport networks based on optical paths," *International Journal of Communication Systems*, Vol. 8, No. 6, Nov.–Dec. 1995, pp. 377–389.

[158] Yasaka, K., H. Ishii, K. Takahata, K. Oe, Y. Yoshikuni, and H. Tsuchiya, "Broad-range tunable wavelength conversion of a 10 Gbit/s signal using a super structure grating distributed Bragg reflector laser," *Tech. Digest, Integrated Photonics Research*, San Francisco, CA, Feb. 17–19, 1994, pp. FC2-1–FC2-3.

[159] Watanabe, A., S. Okamoto, K. Sato, and M. Okuno, "New Optical Path Cross-Connect Architectures Offering High Modularity," *2nd Asia-Pacific Conference on Communications, APCC '95*, Osaka, Japan, June 13–16, 1995, pp. 89–92.

[160] Koga, M., Y. Hamazumi, A. Watanabe, S. Okamoto, H. Obara, K. Sato, S. Suzuki, and M. Okuno, "Design and performance of optical path cross-connect system based on wavelength path concept," *Special Joint Issue of IEEE Journal of Lightwave Technology and IEEE Journal on Selected Areas in Communications*, 1996, Vol. 14, No. 6.

[161] Khoe, G. D., G. Heydt, I. Borges, P. Demeester, A. Ebberg, A. Labrujere, and J. Rawsthorne, "Coherent multicarrier technology for implementation in the customer access," *J. Lightwave Technology*, Vol. 11, No. 5/6, May/June 1993, pp. 695–713.

[162] Ebberg, A., R. Noe, L. Stoll, and R. Schimpe, "A coherent OFDM switching system for flexible optical network configuration," *J. Lightwave Technology*, Vol. 11, No. 5/6, May/June 1993, pp. 847–853.

[163] Kawachi, M., "Planer lightwave circuits for optical FDM," *OEC '92 Tech. Digest*, July 1992, 17C1-1.

[164] Yoshida, J., and M. Kawachi, "Integrated optical circuit technologies," *NTT Review*, Vol. 4, No. 6, Nov. 1992, pp. 92–96.

[165] Kawachi, M., et al., "Silica-based optical-matrix switch with intersecting Mach-Zehnder waveguides for larger fabrication tolerances," *Proc. OFC/IOOC '93*, March 1993, TuH4.

[166] Watanabe, A., S. Okamoto, and K. Sato, "Optical path cross-connect system architecture suitable for large scale expansion," *IEEE Journal of Lightwave Technology*, 1997.

List of Acronyms

A
ADM	add/drop multiplexer
ADP	adapter
APS	automatic protection switching
ASE	amplified spontaneous emission
ATD	asynchronous time division
ATM	asynchronous transfer mode
AU-n	administrative unit-n
AUG	administrative unit group

B
BER	bit error rate
BIP	bit interleaved parity
B-ISDN	broadband integrated services digital network
B-TE	terminal equipment for B-ISDN
B-NT	network termination for B-ISDN

C
CAC	connection admission control
CAD	computer aided design
CBR	constant bit rate
CCITT	Consultative Committee for International Telegraph and Telephone
CDV	cell delay variation
CEQ	customer equipment
CLP	cell loss priority
CPN	customer premises network
C-x	container-x

D
dB	decibel
DCC	data communications channel
DCS	digital cross-connect system
DC-switch	delivery and coupling switch
DMUX	demultiplexer

E
EDFA	erbium-doped-fiber amplifier
E/O	electrical-to-optical

F
FCS	fast circuit switching
FDDI	fiber distributed data interface
FDM	frequency division multiplexing
FPS	fast packet switching
FTTC	fiber to the curb
FTTH	fiber to the home

G
GFC	generic flow control

H
HEC	header error control

I
IEEE	Institute of Electrical and Electronics Engineers
ITU-T	Telecommunication Standardization Sector of International Telecommunication Union

L
LAN	local area network
LB	Leaky Bucket (algorithm)
LD	laser diode
LSI	large-scale integration
LS-IF	low-speed interface
LT	line termination

M
MAC	media access control
MAN	metropolitan area network
MPEG	motion picture experts group
MS	multiplex section
MSOH	multiplex section overhead
MUX	multiplexer
MWSF	multiwavelength tunable filter

N
NE	network element
NLC	normalized link capacity
NNI	network node interface
NPC	network parameter control
NPL	normalized processing load
NT	network termination
NTT	Nippon Telegraph and Telephone

O
OA&M	operation, administration and maintenance
O/E	optical-to-electrical
OFA	optical fiber amplifier
OH	overhead
OLT	optical line terminal
ONU	optical network unit
OPC	optical path connection
OP-XC	optical path cross-connect
OR	optical receiver
OS	optical sender

P
PC	optical power coupler
PDH	plesiochronous digital hierarchy
PDS	passive double star
PDU	protocol data unit
PLC	planar lightwave circuit
POH	path overhead
PON	passive optical network
POTS	plain old telephone service
PT	payload type
PVC	permanent virtual circuit

Q
QOS — quality of service

R
RS — regenerator section
RSOH — regenerator section overhead

S
SAP — service access point
SC — star coupler
SCM — subcarrier multiplexing
SD — space division
SDH — synchronous digital hierarchy
SDM — space division multiplexing
SLT — subscriber line termination
SMDS — switched multimegabit data service
SNR — signal-to-noise ratio
SOH — section overhead
STD — source traffic descriptor
STM — synchronous transfer mode
STM-N — synchronous transport module level N

T
TDM — time division multiplexing
TE — terminal equipment
TP — transmission path
TWF — tunable-wavelength filter

U
UNI — user-network interface
UPC — usage parameter control

V
VBR — variable bit rate
VC — virtual container
VC — virtual channel
VCC — virtual channel connection
VCI — virtual channel Identifier
VC-n — virtual container-n
VDR — virtual direct routing
VP — virtual path
VPC — virtual path connection
VPG — virtual path group
VPI — virtual path identifier
VPT — virtual path termination
VWP — virtual wavelength path

W
W-D	wavelength division demultiplexer
WDM	wavelength division multiplexing
W-DMUX	wavelength division demultiplexer
W-M	wavelength division multiplexer
W-MUX	wavelength division multiplexer
WP	wavelength path
WTF	wavelength-tunable filter

X
XC	cross-connect

About the Author

Ken-ichi Sato received B.S., M.S., and Ph.D. degrees in electronics engineering from the University of Tokyo, Tokyo, Japan, in 1976, 1978, and 1986, respectively. In 1978, he joined Yokosuka Electrical Communication Laboratories, NTT. From 1978 to 1984, he was engaged in the research and development of optical-fiber transmission technologies. His R&D experiences cover fiber-optic video transmission systems for CATV distribution systems and subscriber loop systems. Since 1985, he has been active in the development of broadband ISDN based on ATM techniques. His main research activities include information transport network architectures, variable network architectures with virtual paths, broadband transport technology, ATM network performance analysis, and photonic network technologies. He has authored/coauthored more than a hundred research publications in international journal and conferences. He currently serves as a senior research engineer and leads a research group, the Photonic Network Research Group, in Optical Network Systems Laboratories at NTT.

Dr. Sato was awarded the NTT Research and Development Award in 1991 and the NTT President Award in 1995 for his contribution to ATM network architecture and system technology development in 1995. He received the Young Engineer Award in 1984 and the Excellent Paper Award in 1991, both from the Institute of Electronics, Information and Communication Engineers (IEICE) of Japan. He is a member of the IEICE of JAPAN and a senior member of the IEEE Communications Society.

Index

Access lines, 3
Acrylic topology, 156
Adapter, 134
Adaptive network contol, 11–13
Add/drop multiplexer, 9, 55, 127, 175–76
ADM. *See* Add/drop multiplexer
Administrative unit, 29–32
Administrative unit pointer, 28–29, 31
ADP. *See* Adapter
Amplified spontaneous emission, 177
Amplifiers, 191
APS. *See* Automatic protection switching
ARPANET, 40
ASE. *See* Amplified spontaneous emission
Asynchronous time division, 42–44
Asynchronous transfer mode, 11, 18
    CBR cell traffic, 87–88
    cell-delay distribution, 93–95
    cell loss priority, 91–92
    cell mapping, 32–34
    cell structure, 47–48
    cross-connect system, 93, 105–8
    equivalent terminal, 72–73
    features of, 44–45
    introduction to, 37–38, 40–44
    network architecture, 45–48, 103–8
    network resource management, 68–84, 101–3
    optical paths, 139–42, 148–51
    reserved-mode connection, 101–3
    source traffic descriptor, 71–72
    statistical cell multiplexing, 103–4
    STM-based transport, 38–40
    traffic shaping, 73–75
    usage parameter control, 75–82
    variable bit rate path, 98–101
    virtual path, 51–53, 84–91, 96–98
    virtual path connection, 48–51
Asynchronous transfer mode, compared to STM
    path accommodation, 55–59
    path bandwidth control, 59–66
    path networks, 53–55
    subscriber networking, 67–68
    transport node architecture, 53–55
ATD. *See* Asynchronous time division
ATM. *See* Asynchronous transfer mode
AU. *See* Administrative unit
AU pointer. *See* Administrative unit pointer
Automatic protection switching, 127

B-ISDN. *See* Broadband-integrated services digital network
Bit rate
    asynchronous transfer mode, 38
    path layers, 22
    path network, 10
    transmission system, 6
Block frame, 26–27
Broadband-integrated services digital network, 37, 196
    service expansion, 127–28
    transmission cost, 121–23
    transport technology advances, 120–21
Buffer size, 93
Burst-level sharing, 82
Burst switching, 41–42

CAC. *See* Connection admission control
CAD. *See* Computer-aided design
Call blocking, 61, 63, 70, 155, 159, 160
Call-by-call processing, 155, 159

Call routing. *See* Dynamic call routing
CBR. *See* Constant bit rate
CCITT. *See* Consultative Committee for International Telegraph and Telephone
CDV. *See* Cell-delay variation
Cell, 38
Cell-based ATM transport network, 134
Cell-delay distribution, 93–98
Cell-delay variation, 73–74, 81–82, 86–91
Cell-level sharing, 83
Cell loss priority, 48, 91–91
Cell/packet transmission, 139–42
Cell structure, ATM, 47–49
Cell traffic modeling, 87–88
CEQ. *See* Customer equipment
Chain topology, 148–49
Child pattern, Leaky Bucket, 77, 79–80
Chooser node, 66
Circuit layer control, 13
Clos switching network, 178–79, 191
CLP. *See* Cell loss priority
Common bus switch, 107
Communication circuit layer, 8
Computer-aided design, 193
Concatenation, administration unit, 31–32
Connecting point, 45–47
Connection, 45
Connection admission control, 96–98
Connection endpoint, 45, 47
Constant bit rate, 51, 84–98
Consultative Committee for International Telegraph and Telephone, 18, 41, 44, 51–52
Container, 26
Controlled transmission, 48
Cost
   B-ISDN, 131, 133–34
   cross-connection, 6, 10, 189–190
   dynamic path layer control, 13
   fiber optics, 38
   STM and ATM, 56, 58–59
   transport network, 13–15, 194–98
   video transmission, 121–23
   wavelength path network, 175
Coupling loss, 183
CPN. *See* Customer premises network
Credit Window method, 75–76
Criss-cross network, 142, 144
Cross-bar network, 142, 144
Cross-connect buffer, 93

Cross-connection
   functions of, 4–7
   generic architecture, 168, 175–177
   hardware cost, 189–190
   node throughput, 131, 132
   optical path requirements, 185–194
   path accommodation, 10–11
   switch examples, 177–185
   synchronous digital hierarchy, 22–23
   virtual path, 53, 105–8
   VPI/VCI, 47
Cross-point buffer switch, 106–7
Crosstalk, 177
Customer equipment, 50
Customer premises network, 103

DACS 1, 126
Data communication channel, 28
DCC. *See* Data communication channel
DCS. *See* Digital cross-connect system
DC switch. *See* Delivery and coupling switch
de Bruijn Graph, 142
Delivery and coupling switch, 181–85, 188–89, 193–94
Demultiplexer, 139
Differential frequency generation, 176
Digital cross-connect system, 120
Digital section level, 45
Digital transport technology, 17–19
Dijkstra's algorithm, 137
Dynamic call routing, 13
Dynamic path network, 60
Dynamic path routing, 14

EDFA. *See* Erbium doped fiber amplifier
Electrical-to-optical circuit, 177
ELI. *See* Existing lower-speed interface
Endpoint, 45, 47, 160
End-to-end cell-delay distribution, 93–95
End-to-end connection, 8, 118–19
E/O circuit. *See* Electrical-to-optical circuit
Equivalent terminal, 72–73
Erbium doped fiber amplifier, 125
Erlang lost-call formula, 3–4
Existing lower-speed interface, 25

Fallback on path application, 64
Fast circuit switching, 41
Fast packet switching, 41
FCS. *See* Fast circuit switching
FDDI. *See* Fiber distributed data interface
FDM. *See* Frequency division multiplexing
Fiber distributed data interface, 119

Fiber optics, 38
Fibers, wavelength path, 167, 169–70, 172–73, 175
Fiber to the curb, 121
Fiber to the home, 121
Filters, wavelength, 180–83
Fixed delay, 86
Fixed-wavelength laser diode, 178
Fixed-wavelength transmitter, 139
Flexibility, network, 7, 9, 11, 117, 131
Four-wave mixing, 176
FPS. *See* Fast packet switching
Frequency division multiplexing, 5, 125, 129
FTTC. *See* Fiber to the curb
FTTH. *See* Fiber to the home

Generic flow control, 48
GFC. *See* Generic flow control
Group transit switching, 13

Header, 38
Header error control, 48
HEC. *See* Header error control
Hierarchical path network, 9–10, 12
Hockham, G. A., 117
Hypercube network, 142, 144

Input buffer, 106–7
Integrated services digital network, 37
Interface structure, ATM, 44
International Telecommunication Union, 31
ISDN. *See* Integrated services digital network
ITU-T. *See* International Telecommunication Union

Japan, digital hierarchy of, 55–57
Jitter, 88, 90–91, 94

Kao, K. C., 117
Kapron, F. P., 117

LAN. *See* Local area network
Large-scale integration, 38
Laser diode, 178
Layered architecture, 7–10
LB method. *See* Leaky bucket method
LD. *See* Laser diode
Leaky Bucket method, 75–81
Link, 45
Link-by-link switching, 5–6
Link capacity sharing, 60–61
Link utilization
    constant bit rate path, 87, 92
    plesiochronous digital hierarchy, 26

STM and ATM, 55–56, 58–59, 63
synchronous digital hierarchy, 23
synchronous transfer mode, 40
variable bit rate path, 86
Local area network, 119
Logical connections, 142, 144
Loss, cross-connect switch, 191–92
Lost-call formula, 3–4
LSI. *See* Large-scale integration

MAC. *See* Media access control
Mach-Zehnder interferometer, 193
Manhattan Street network, 142, 144
Mapping, cell, 32–34
Matrix switch, 106
M/D/1 model, 88–89, 91
Media access control, 82
Modular growth, switch, 186–90
Motion Picture Experts Group, 121
MPEG. *See* Motion Picture Experts Group
MS. *See* Multiplex section
MSOH. *See* Multiplex section overhead
Multihop network, 139, 141–45, 148–51
Multiple-fiber link, 166–75
Multiplexer, 23
Multiplexing
    administrative unit, 29–32
    asynchronous time division, 42–44
    asynchronous transfer mode, 32–34, 68–69
    cross-connection, 10
    nonhierarchical, 44–45
    packet, 41
    plesiochronous digital hierarchy, 23
    section overhead, 28–29
    statistical gain, 103–4
    synchronous transport mode, 26–28, 68–69
    VPs experiencing CDV, 88–91
Multiplex section, 20
Multiplex section overhead, 28
Multiwavelength selective filter, 180–82
MUX. *See* Multiplexer
MWSF. *See* Multiwavelength selective filter

NE. *See* Network element
Network element, 103–5
    cross-connect system, 105–8
    other, 108–9
Network layering, 45–47
Network-network application, 48, 50, 84, 86
Network node interface, 17–18, 22–26, 37, 47–49
Network parameter control, 73, 75

Network resource management, ATM, 68–70
   CBR path accommodation, 84–98
   princple of, 70–82
   sharing principle, 82–84
   utilization enhancement, 101–3
   VBR path accommodation, 98–101
Network restoration, 7, 9
   with multiple-fiber links, 166–75
   with optical paths, 136–38, 160–62
   with single-fiber links, 162–66
   technology advancement, 127
Network termination, 108–9
Nippon Telegraph and Telephone, 23, 51, 123–26
NLC. *See* Normalized link capacity
NNI. *See* Network node interface
Nodes
   architecture, 53–55
   cross-connect, 10, 131–32
   failed, 160
   technology advancement, 125–26
   in WP/VWP models, 169, 172
Nonblocking switch, 185–86
Nonhierarchical network, 12
Normalized link capacity, 61
Normalized processing load, 63
NPC. *See* Network parameter control
NPL. *See* Normalized processing load
NT. *See* Network termination
NTT. *See* Nippon Telegraph and Telephone

OA&M. *See* Operation, administration, and maintenance
O/E circuit. *See* Optical-to-electrical circuit
OFA. *See* Optical-fiber amplifier
OH. *See* Overhead
OPC. *See* Optical path connection
Operation, administration, and maintenance, 8, 22, 28, 138
Optical-fiber amplifier, 198
Optical-fiber transmission, 5
Optical path, 139
   cross-connect architecture, 175–77
   cross-connect requirements, 185–94
   cross-connect switches, 177–85
   network cost, 194–98
   network restoration, 160–62
   realization technologies, 138–54
   technology benefits, 130–38
   transport network photonization, 129
   WDM benefits, 129–30
Optical receiver, 183

Optical regeneration, 177, 183
Optical sender, 178, 183
Optical-to-electrical circuit, 176–77
Optical transmission
   advancement in, 123–25
   history of, 117–18
   in networks, 118–20
OR. *See* Optical receiver
OS. *See* Optical sender
Output buffer, 106–7
Overhead, 22, 28–30

Packet switching, 40–41
Passive double star, 67, 118
Passive optical network, 67, 118
Path, 5, 41, 52
   See also Optical path; Virtual wavelength path; Wavelength path
Path accommodation, 10–12
   ATM and STM, 55–59
   constant bit rate, 84–98
   variable bit rate, 86, 98–101
Path bandwidth control, 59–66
Path cross-connection, 5–7
Path network, 6
   ATM and STM, 53–55
   layers of, 8–10, 13, 20–23
Path overhead, 20
Path route optimization, 156–57
Payload, 38
Payload type, 48
PDH. *See* Plesiochronous digital hierarchy
PDS. *See* Passive double star
Perfect shuffle network, 142, 144
Photonization, transport network, 129
Physical layer, 8, 45
Plain old telephone service, 38
Planar lightwave circuit, 186, 191–93
PLC. *See* Planar lightwave circuit
Plesiochronous digital hierarchy, 5
   cross-connect system, 105–6
   failure restoration, 136
   features of, 17–19
   node architecture, 24, 26, 125–26
   path stages, 23
   technology, 134
POH. *See* Path overhead
Point-to-multipoint connection, 146
Point-to-point connection, 118–19, 148
Poisson traffic, 90–91, 98
PON. *See* Passive optical network
Ports, number of, 166–67, 169, 171

Postfix network, 176–78, 181
POTS. *See* Plain old telephone service
Power splitter, 180–81
Prefix network, 176, 179
PRELUDE, 42
Propagation delay, 86
PT. *See* Payload type

QOS. *See* Quality of service
Quality of service, 70–71, 77, 83, 86, 92, 98, 126

Rate adaptation, ATM, 44
Reference configuration, 72–73
Regenerator section, 20
Regenerator section level, 45
Regenerator section overhead, 28
Repeaters, 177
Reserved-mode connection, 101–3
Resource allocation, 68
Resource sharing, 82–84
Restoration, virtual path, 64–66
Ring topology, 148, 149
RS. *See* Regenerator section
RSOH. *See* Regenerator section overhead

SAP. *See* Service access point
SCM. *See* Subcarrier multiplexing
SDH. *See* Synchronous digital hierarchy
SDM. *See* Space division multiplexing
SD switch. *See* Space division switch
Section overhead, 28–30
Self-healing, virtual path, 65–66
Semiconductor laser, 118
Service access, 6
Service access point, 72
Service network layer, 8
Shared-buffer switch, 106–7
Signal-to-noise ratio, 191
Single-fiber link, 156–60, 162–66
Single-hop network, 139, 141–43, 147
Single-mode fiber, 25
Single star architecture, 57
Sliding Window method, 75–76
SLT. *See* Subscriber line termination
SMDS. *See* Switched multimegabit data service
SMF. *See* Single-mode fiber
SNR. *See* Signal-to-noise ratio
SOH. *See* Section overhead
Source traffic descriptor, 71–72
Space division multiplexing, 129

Space division switch, 177–80, 186–90, 193–94
Star coupler, 139, 142, 146, 148, 181, 183
Statistical cell multiplexing gain, 103–4
STM. *See* Synchronous transfer mode
STM-N. *See* Synchronous transport module-N
Subcarrier multiplexing, 134
Subscriber line termination, 93, 103
Subscriber networking, 67–68
Supercube network, 142, 144
Switch architecture, ATM, 106–7
Switched communications networks, 1–4
Switched multimegabit data service, 120
Switching
   cross-connect loss, 191–92
   delivery and coupling, 181–85, 188–89, 193–94
   group transit, 13
   modular growth, 186–90
   nonblocking, 185–86
   space division, 177–80, 186–90, 193–94
   using power splitters, 180–81
Synchronous digital hierarchy, 6, 8, 10, 56, 58, 126, 134, 136
   advantages of, 18–19
   cross-connect system, 105
   multiplexing principles, 26–34
   network cost, 194–96
   network node interface, 22–26
   transport network layers, 20–22
Synchronous transfer mode
   path accommodation, 55–59
   path bandwidth control, 59–66
   path networks, 53–55
   subscriber networking, 67–68
   transport techniques, 38–40
Synchronous transport module-N, 26–28

T1 span, 42
Tandem delay, 94–95, 97
TDM. *See* Time division multiplexing
Time division multiplexing, 39, 42, 53, 59, 129–30, 197
Traffic shaping, 73
Transmission capacity, optical path, 130
Transmission facilities network, 12
Transmission media network, 10–11
Transmission path layer, 8
Transmission path level, 45
Transport network
   adaptive controls, 11–13
   cost, 13–15

Transport network (continued)
  cross-connect functions, 4–7
  layered architecture, 7–10, 20–22
  path accommodation, 10–11
  switched communications, 1–4
Transport node. *See* Nodes
Transport technology
  advances in, 120–21
  B-ISDN, 127–28
  network restoration, 127
  node advancement, 125–26
  optical system, 123–25
Tributary signals, 26–27, 31
Trunks, 3–5, 118
Tunable receiver, 139
Tunable wavelength filter, 180–81, 183
Tunable-wavelength laser diode, 178
TWF. *See* Tunable wavelength filter

UNI. *See* User network interface
Unidirectional link, 47
UPC. *See* Usage parameter control
Usage parameter control, 71, 73, 75–83, 99–101, 103
User-network application, 37, 47–50, 84, 130
User-user application, 48, 50, 84

Variable bit rate, 51, 70, 86, 98–101
Variable commuications network, 13
Variable delay, 86
VBR. *See* Variable bit rate
VC. *See* Virtual channel; Virtual container
VCC. *See* Virtual channel connection
VCI. *See* Virtual channel identifier
VDR. *See* Virtual direct routing
Video transmission, 121–23
Virtual channel, 45–46, 52, 70
Virtual channel connection, 47
Virtual channel identifier, 46, 48
Virtual container, 20, 22, 26, 28–29, 31–34
Virtual container-n, 31
Virtual direct routing, 51
Virtual path, 37
  bandwidth control, 60–63
  capacity of, 60
  CBR design, 84–98

  channels, 48, 50–51
  as fallback protection, 64
  in network layer, 45–47
  self-healing, 65–66
  technology evolution, 51–53, 126
Virtual path connection, 46–47
Virtual path group, 136
Virtual path identifier, 47–48, 53, 64, 146
Virtual path terminator, 51, 53, 64
Virtual wavelength path, 142, 144–48
  compared to multihop, 148–51
  compared to wavelength path, 151–54
  design with network restoration, 160–75
  design without network restoration, 155–6
  evolution from WP to, 190–91
Voice communication, compared to video, 12
VP. *See* Virtual path
VPC. *See* Virtual path connection
VPG. *See* Virtual path group
VPI. *See* Virtual path identifier
VPT. *See* Virtual path terminator
VWP. *See* Virtual wavelength path

Wavelength assignment, 156–59
Wavelength conversion, 175–77
Wavelength division multiplexing, 125, 129–30, 196
Wavelength filters, 180–83
Wavelength number
  limited, 159–60, 166–75
  unlimited, 156–59, 162–66
Wavelength path, 142, 144–48
  compared to multihop, 148–51
  compared to VWP, 151–54
  design with network restoration, 160–75
  design without network restoration, 155–6
  evolution to VWP, 190–91
Wavelength reuse, 151
WDM. *See* Wavelength division multiplexing
Worst UPC output pattern, 99–101
WP. *See* Wavelength path

X.25 standard, 41

The Artech House Telecommunications Library

Vinton G. Cerf, Series Editor

Advanced Technology for Road Transport: IVHS and ATT, Ian Catling, editor

Advances in Computer Communications and Networking, Wesley W. Chu, editor

Advances in Computer Systems Security, Rein Turn, editor

Advances in Telecommunications Networks, William S. Lee and Derrick C. Brown

Advances in Transport Network Technologies: Photonics Networks, ATM, and SDH, Ken-ichi Sato

Analysis and Synthesis of Logic Systems, Daniel Mange

An Introduction to International Telecommunications Law, Charles H. Kennedy and M. Veronica Pastor

An Introduction to U.S. Telecommunications Law, Charles H. Kennedy

Asynchronous Transfer Mode Networks: Performance Issues, Raif O. Onvural

ATM Switching Systems, Thomas M. Chen and Stephen S. Liu

A Bibliography of Telecommunications and Socio-Economic Development, Heather E. Hudson

Broadband: Business Services, Technologies, and Strategic Impact, David Wright

Broadband Network Analysis and Design, Daniel Minoli

Broadband Telecommunications Technology, Byeong Lee, Minho Kang, and Jonghee Lee

Cellular Radio: Analog and Digital Systems, Asha Mehrotra

Cellular Radio Systems, D. M. Balston and R. C. V. Macario, editors

Client/Server Computing: Architecture, Applications, and Distributed Systems Management, Bruce Elbert and Bobby Martyna

Codes for Error Control and Synchronization, Djimitri Wiggert

Communications Directory, Manus Egan, editor

The Complete Guide to Buying a Telephone System, Paul Daubitz

Computer Networks: Architecture, Protocols, and Software, John Y. Hsu

Computer Telephone Integration, Rob Walters

The Corporate Cabling Guide, Mark W. McElroy

Corporate Networks: The Strategic Use of Telecommunications, Thomas Valovic

Current Advances in LANs, MANs, and ISDN, B. G. Kim, editor

Digital Beamforming in Wireless Commuunications, John Litva, Titus Kwok-Yeung Lo

Digital Cellular Radio, George Calhoun

Digital Hardware Testing: Transistor-Level Fault Modeling and Testing, Rochit Rajsuman, editor

Digital Signal Processing, Murat Kunt

Digital Switching Control Architectures, Giuseppe Fantauzzi

Distributed Multimedia Through Broadband Communications Services, Daniel Minoli and Robert Keinath

Disaster Recovery Planning for Telecommunications, Leo A. Wrobel

Distance Learning Technology and Applications, Daniel Minoli

Document Imaging Systems: Technology and Applications, Nathan J. Muller

EDI Security, Control, and Audit, Albert J. Marcella and Sally Chen

Electronic Mail, Jacob Palme

Enterprise Networking: Fractional T1 to SONET, Frame Relay to BISDN, Daniel Minoli

Expert Systems Applications in Integrated Network Management, E. C. Ericson, L. T. Ericson, and D. Minoli, editors

FAX: Digital Facsimile Technology and Applications, Second Edition, Dennis Bodson, Kenneth McConnell, and Richard Schaphorst

FDDI and FDDI-II: Architecture, Protocols, and Performance, Bernhard Albert and Anura P. Jayasumana

Fiber Network Service Survivability, Tsong-Ho Wu

Fiber Optics and CATV Business Strategy, Robert K. Yates et al.

A Guide to Fractional T1, J. E. Trulove

A Guide to the TCP/IP Protocol Suite, Floyd Wilder

Implementing EDI, Mike Hendry

Implementing X.400 and X.500: The PP and QUIPU Systems, Steve Kille

Inbound Call Centers: Design, Implementation, and Management, Robert A. Gable

Information Superhighways: The Economics of Advanced Public Communication Networks, Bruce Egan

Integrated Broadband Networks, Amit Bhargava

Intelcom '94: The Outlook for Mediterranean Communications, Stephen McClelland, editor

International Telecommunications Management, Bruce R. Elbert

International Telecommunication Standards Organizations, Andrew Macpherson

Internetworking LANs: Operation, Design, and Management, Robert Davidson and Nathan Muller

Introduction to Document Image Processing Techniques, Ronald G. Matteson

Introduction to Error-Correcting Codes, Michael Purser

Introduction to Satellite Communication, Bruce R. Elbert

Introduction to T1/T3 Networking, Regis J. (Bud) Bates

Introduction to Telecommunication Electronics, Second Edition, A. Michael Noll

Introduction to Telephones and Telephone Systems, Second Edition, A. Michael Noll

Introduction to X.400, Cemil Betanov

Land-Mobile Radio System Engineering, Garry C. Hess

LAN/WAN Optimization Techniques, Harrell Van Norman

LANs to WANs: Network Management in the 1990s, Nathan J. Muller and Robert P. Davidson

Long Distance Services: A Buyer's Guide, Daniel D. Briere

Measurement of Optical Fibers and Devices, G. Cancellieri and U. Ravaioli

Meteor Burst Communication, Jacob Z. Schanker

Minimum Risk Strategy for Acquiring Communications Equipment and Services, Nathan J. Muller

Mobile Communications in the U.S. and Europe: Regulation, Technology, and Markets, Michael Paetsch

Mobile Information Systems, John Walker

Narrowband Land-Mobile Radio Networks, Jean-Paul Linnartz

Networking Strategies for Information Technology, Bruce Elbert

Numerical Analysis of Linear Networks and Systems, Hermann Kremer et al.

Optimization of Digital Transmission Systems, K. Trondle and Gunter Soder

Packet Switching Evolution from Narrowband to Broadband ISDN, M. Smouts

Packet Video: Modeling and Signal Processing, Naohisa Ohta

Personal Communication Systems and Technologies, John Gardiner and Barry West, editors
The PP and QUIPU Implementation of X.400 and X.500, Stephen Kille
Practical Computer Network Security, Mike Hendry
Principles of Secure Communication Systems, Second Edition, Don J. Torrieri
Principles of Signaling for Cell Relay and Frame Relay, Daniel Minoli and George Dobrowski
Principles of Signals and Systems: Deterministic Signals, B. Picinbono
Private Telecommunication Networks, Bruce Elbert
Radio-Relay Systems, Anton A. Huurdeman
Radiodetermination Satellite Services and Standards, Martin Rothblatt
Residential Fiber Optic Networks: An Engineering and Economic Analysis, David Reed
Secure Data Networking, Michael Purser
Service Management in Computing and Telecommunications, Richard Hallows
Setting Global Telecommunication Standards: The Stakes, The Players, and The Process, Gerd Wallenstein
Smart Cards, José Manuel Otón and José Luis Zoreda
Super-High-Definition Images: Beyond HDTV, Naohisa Ohta, Sadayasu Ono, and Tomonori Aoyama
Television Technology: Fundamentals and Future Prospects, A. Michael Noll
Telecommunications Technology Handbook, Daniel Minoli
Telecommuting, Osman Eldib and Daniel Minoli
Telemetry Systems Design, Frank Carden
Telephone Company and Cable Television Competition, Stuart N. Brotman
Teletraffic Technologies in ATM Networks, Hiroshi Saito
Terrestrial Digital Microwave Communications, Ferdo Ivanek, editor
Toll-Free Services: A Complete Guide to Design, Implementation, and Management, Robert A. Gable
Transmission Networking: SONET and the SDH, Mike Sexton and Andy Reid
Transmission Performance of Evolving Telecommunications Networks, John Gruber and Godfrey Williams
Troposcatter Radio Links, G. Roda
Understanding Emerging Network Services, Pricing, and Regulation, Leo A. Wrobel and Eddie M. Pope
Understanding Networking Technology: Concepts, Terms and Trends, Mark Norris

UNIX Internetworking, Uday O. Pabrai

Videoconferencing and Videotelephony: Technology and Standards, Richard Schaphorst

Virtual Networks: A Buyer's Guide, Daniel D. Briere

Voice Processing, Second Edition, Walt Tetschner

Voice Teletraffic System Engineering, James R. Boucher

Wireless Access and the Local Telephone Network, George Calhoun

Wireless Data Networking, Nathan J. Muller

Wireless LAN Systems, A. Santamaría and F. J. López-Hernández

Wireless: The Revolution in Personal Telecommunications, Ira Brodsky

Writing Disaster Recovery Plans for Telecommunications Networks and LANs, Leo A. Wrobel

X Window System User's Guide, Uday O. Pabrai

For further information on these and other Artech House titles, contact:

Artech House
685 Canton Street
Norwood, MA 02062
617-769-9750
Fax: 617-769-6334
Telex: 951-659
e-mail: artech@artech-house.com

Artech House
Portland House, Stag Place
London SW1E 5XA England
+44 (0) 171-973-8077
Fax: +44 (0) 171-630-0166
Telex: 951-659
e-mail: artech-uk@artech-house.com